The ASQ CSQP Study Guide

THE ASQ CSQP STUDY GUIDE

Mark Allen Durivage
Edward Cook

ASQ Quality Press
Milwaukee, Wisconsin

American Society for Quality, Quality Press, Milwaukee 53203
© 2018 by ASQ
All rights reserved. Published 2018

26 25 24 23 LS 8 7 6

Library of Congress Cataloging-in-Publication Data

Names: Durivage, Mark Allen, author. | Cook, Edward, author.
Title: The ASQ CSQP study guide / Mark Allen Durivage, Edward Cook.
Description: Milwaukee, Wis. : ASQ Quality Press, [2018] | Includes
    bibliographical references.
Identifiers: LCCN 2018014554 | ISBN 9780873899703 (soft cover, spiral bound :
    alk. paper) ; ISBN 9781636941295 (paperback)

Subjects: LCSH: Acceptance sampling. | Quality control.
Classification: LCC TS156.4 D86 2018 | DDC 658.4/013—dc23
LC record available at https://lccn.loc.gov/2018014554

ASQ advances individual, organizational, and community excellence worldwide through learning, quality improvement, and knowledge exchange.

Bookstores, wholesalers, schools, libraries, and organizations: Quality Press books are available at quantity discounts with bulk purchases for business, trade, or educational uses. For more information, please contact Quality Press at 800-248-1946 or books@asq.org.

To place orders or browse the full selection of Quality Press titles, visit our website at:
http://www.asq.org/quality-press.

Quality Press
600 N. Plankinton Ave.
Milwaukee, WI 53203-2914
Email: books@asq.org

ASQ Excellence Through Quality®

I would like to dedicate this book to and recognize the patience of my wife, Dawn, and my sons, Jack and Sam, who allowed me time to complete this project. I would also like to thank Edward Cook for partnering on this project, and in the process becoming my friend.

—Mark Allen Durivage, ASQ Fellow

I would like to dedicate this work to my children (Ryan, Grace, Nick, and Aisha) as an example of both the fruit borne from hard work, and our obligation to give back and share knowledge. I would also like to thank Mark Durivage for inviting me to cowrite with him, and for his friendship and support.

—Edward G. Cook

Table of Contents

Introduction

This book is primarily meant to aid those taking the ASQ Certified Supplier Quality Professional (CSQP) exam, and is best used in conjunction with *The Certified Supplier Quality Professional Handbook* (ASQ Quality Press). Section 1 provides 336 practice questions organized by the seven parts of the 2016 Body of Knowledge (BoK). Section 2 gives the reader a 150-question practice exam comprising each of the seven parts of the BoK, in a randomized order, that simulates the actual certification exam.

Unlike other resources on the market, all these questions and solutions were developed specifically to address the 2016 CSQP Body of Knowledge and help those studying for the certification, including considering the proper depth of knowledge and required levels of cognition.

Please note that all calculations for these questions were performed using a simple scientific calculator. Therefore, some answers may vary slightly if worked with a spreadsheet or other software application. The authors strongly suggest referring to the official ASQ calculator requirements,* purchasing a compliant calculator, and practicing using the calculator. Additionally, we suggest purchasing *The Certified Supplier Quality Professional Handbook* from ASQ Quality Press and becoming familiar with the equations and statistical tables provided in the appendixes to optimize your time during the actual certification exam.

As a reminder, practice/sample test questions such as those in this study guide cannot be taken into ASQ certification exam rooms. The certification exam is open book. It is highly recommended that you do take the *Certified Supplier Quality Professional Handbook* in with you to look up or verify any answers as you work the exam questions.

We welcome your feedback and suggestions for improvement. Please contact us at books@asq.org, and we will do our best to clarify any questions you may have and incorporate any suggestions for improvement into future printings or editions of this study guide.

Mark Allen Durivage
Lambertville, Michigan

Edward Cook
Chicago, Illinois

---

*https://asq.org/cert/faq/calculator

Acknowledgments

The *ASQ CSQP Study Guide* is dedicated to the hardworking individuals globally who work tirelessly for their companies to improve and optimize service and production processes.

We would like to thank those who have inspired, taught, and trained us throughout our professional careers. Additionally, we would like to thank ASQ Quality Press, especially Paul O'Mara, Managing Editor, and Randy Benson, Senior Creative Services Specialist, for their expertise and technical competence, which made this project a reality.

LIMIT OF LIABILITY/DISCLAIMER OF WARRANTY

The authors have put forth their best effort in compiling the content of this book; however, no warranty with respect to the material's accuracy or completeness is made. Additionally, no warranty is made in regard to applying the recommendations made in this book to any business structure or environments. Businesses should consult regulatory, quality, and/or legal professionals prior to deciding on the appropriateness of advice and recommendations made within this book. The authors shall not be held liable for loss of profit or other commercial damages resulting from the employment of recommendations made within this book, including special, incidental, consequential, or other damages.

Section 1
Practice Questions

This section is divided into seven parts, one for each section in the Certified Supplier Quality Professional (CSQP) Body of Knowledge (BoK). In each part there is a set of questions followed by detailed solutions.

Part I

Supplier Strategy

(54 questions)

A. SUPPLY CHAIN VISION/MISSION

Assist in the development and communication of the supply chain vision/mission statement. (Apply)

B. SUPPLIER LIFECYCLE MANAGEMENT

1. *Supplier selection.* Develop the process for supplier selection and qualification including the identification of sub-tier suppliers, using tools such as SIPOC and decision analysis. (Create)

2. *Performance monitoring.* Develop the supplier performance monitoring system including; expected levels of performance, process reviews, performance evaluations, improvement plans, and exit strategies. (Create)

3. *Supplier classification system.* Define a supplier classification system, e.g. non-approved, approved, preferred, certified, partnership, and disqualified. (Create)

4. *Partnerships and alliances.* Identify and analyze strategies for developing customer–supplier partnerships and alliances. (Analyze)

C. SUPPLY CHAIN COST ANALYSIS

1. *Cost reduction.* Identify and apply relevant inputs to prioritize cost reduction opportunities. (Analyze)

2. *Supply chain rationalization.* Interpret and analyze the optimization of a supply base to improve spending and leverage investments into supplier quality, or risk reduction. (Analyze)

3. *Make/buy decisions.* Provide input on make/buy decisions by using internal and external capability analysis. Apply tools such as, SWOT analysis and use historical performance to analyze requirements. (Analyze)

D. SUPPLIER AGREEMENTS OR CONTRACTS

Review and provide input for developing terms and conditions that govern supplier relationships to ensure quality considerations are addressed. (Apply)

E. DEPLOYMENT OF STRATEGY AND EXPECTATIONS

Communicate strategy internally, and communicate expectations to suppliers externally. (Apply)

QUESTIONS

1. The primary difference between mission and vision statements is:

   a. mission statements relate to the future of the company.

   b. vision statements dictate the future state of the company.

   c. mission statements explain the company's reason for existence.

   d. mission and vision statements are essentially the same and are interchangeable.

2. Which statement is an example of a vision statement?

   a. To procure the highest-quality components from the most reliable suppliers at the best cost possible.

   b. To comply with all applicable standards and regulations.

   c. To complete all supplier audits per the specified schedule.

   d. To create a reliable supply chain through unwavering commitment to quality and cost.

3. Of the following organizational concepts, which would be best associated with supporting metrics?

   a. Corporate purpose

   b. Vision statement

   c. Mission statement

   d. Objectives (strategic and tactical)

4. With the globalization of the supply chain, supplier partnerships are increasingly necessary to:

   a. comply with regulations.

   b. deliver quality products.

   c. ensure market share.

   d. provide export licenses.

5. A commitment and trust are elements of:

   a. partnering.

   b. sourcing.

   c. development.

   d. improvement.

6. Which of the following models is the best for identifying stakeholders of a proposed change in the supply chain?

   a. SIPOC

   b. PDCA

   c. DMAIC

   d. DFSS

7. The SLM model is an integrated approach that considers business and quality needs of the organization. SLM is:

   a. supplier learning model.

   b. supplier life cycle management.

   c. supplier leverage model.

   d. supplier lean method.

8. The supply chain management (SCM) model focuses more on:

   a. supplier integration and costs.

   b. supplier development.

   c. supplier change management.

   d. supplier franchising opportunities.

9. Which of the following models is the best for identifying stakeholders of a proposed change in a manufacturing process?

   a. PDCA

   b. DMAIC

   c. SIPOC

   d. DFSS

10. The primary difference between supplier life cycle management (SLM) and supply chain management (SCM) is:

   a. SLM focuses on supplier integration.

   b. SLM's assessment of suppliers' assets and capabilities.

c. SCM's planning for turnover in suppliers.

d. SCM focuses on supplier integration.

11. Calculate the performance index (PI) for a purchased part that costs $1000.00 and associated unproductive costs of $100.

    a. 1.0

    b. 1.1

    c. 100

    d. 1000

Use the Table below to answer questions 12–15.

	Supplier A	Supplier B	Supplier C	Supplier D
Part cost	$1750	$1650	$1835	$1800
Performance index (PI)	1.3	1.4	1.1	1.2

12. Which supplier will deliver the lowest elevated cost per part?

    a. Supplier A

    b. Supplier B

    c. Supplier C

    d. Supplier D

13. Which supplier will deliver the highest elevated cost per part?

    a. Supplier A

    b. Supplier B

    c. Supplier C

    d. Supplier D

14. What is the elevated cost for Supplier A?

    a. $2018.50

    b. $2160.00

    c. $2275.00

    d. $2310.00

15. What is the elevated cost for supplier B?

    a. $2018.50

    b. $2160.00

    c. $2275.00

    d. $2310.00

16. Strategic products are characterized as products with:

    a. low profit impact and low supply risk.

    b. low profit impact and high supply risk.

    c. high profit impact and low supply risk.

    d. high profit impact and high supply risk.

17. Leverage products are characterized as products with:

    a. low profit impact and low supply risk.

    b. low profit impact and high supply risk.

    c. high profit impact and low supply risk.

    d. high profit impact and high supply risk.

18. Bottleneck products are characterized as products with:

    a. low profit impact and low supply risk.

    b. low profit impact and high supply risk.

    c. high profit impact and low supply risk.

    d. high profit impact and high supply risk.

19. Routine products are characterized as products with:

    a. low profit impact and low supply risk.

    b. low profit impact and high supply risk.

    c. high profit impact and low supply risk.

    d. high profit impact and high supply risk.

20. According to the Kraljic portfolio segmentation model, a product or service with low profit impact, high supply risk, and high sourcing difficulty would be considered:

    a. leverage products.

   b. strategic products.

   c. routine products.

   d. bottleneck products.

21. Strategic products deserve the most attention from purchasing managers. Options include developing long-term supply relationships, regularly analyzing and managing risks, planning for contingencies, and producing the item in-house rather than buying it, if appropriate. Strategic products are considered:

   a. low profit impact, high supply risk.

   b. low profit impact, low supply risk.

   c. high profit impact, high supply risk.

   d. high profit impact, low supply risk.

22. Purchasing approaches for routine products include using standardized products, monitoring and/or optimizing order volume, and optimizing inventory levels. Routine products are considered:

   a. low profit impact, high supply risk.

   b. low profit impact, low supply risk.

   c. high profit impact, high supply risk.

   d. high profit impact, low supply risk.

23. Purchasing approaches to consider for leverage products include using your full purchasing power, substituting products or suppliers, and placing high-volume orders. Leverage products are considered:

   a. low profit impact, high supply risk.

   b. low profit impact, low supply risk.

   c. high profit impact, high supply risk.

   d. high profit impact, low supply risk.

24. Useful approaches for bottleneck products include over-ordering when the item is available (lack of reliable availability is one of the most common reasons that supply is disrupted) and looking for ways to control vendors. Bottleneck products are considered:

   a. low profit impact, high supply risk.

   b. low profit impact, low supply risk.

   c. high profit impact, high supply risk.

   d. high profit impact, low supply risk.

25. Which of the following tools can help in understanding the current process so that improvements can be made?

   a. FMEA

   b. Pareto chart

   c. DMAIC

   d. SIPOC

26. During which stage of the PDCA cycle would project standardization be applied to all other suppliers?

   a. Plan

   b. Do

   c. Check

   d. Act

27. During which stage of the PDCA cycle would supply chain management program development and procurement activities be started with a few selected suppliers?

   a. Plan

   b. Do

   c. Check

   d. Act

28. During which stage of the PDCA cycle would a variety of tools such as benchmarking and supplier scorecards be utilized?

   a. Plan

   b. Do

   c. Check

   d. Act

29. Which stage of the PDCA cycle would start with a spend analysis and diagnosis of the value being purchased?

   a. Plan

   b. Do

   c. Check

   d. Act

30. Costs associated with customer complaints are best characterized as:

   a. appraisal costs.

   b. prevention costs.

   c. internal failure costs.

   d. external failure costs.

31. The costs associated with acceptance sampling are best characterized as:

   a. appraisal costs.

   b. prevention costs.

   c. internal failure costs.

   d. external failure costs.

32. The costs associated with the operation and activities of the material review board (MRB) are best characterized as:

   a. appraisal costs.

   b. prevention costs.

   c. internal failure costs.

   d. external failure costs.

33. The costs associated with the implementation of a companywide training initiative are best characterized as:

   a. appraisal costs.

   b. prevention costs.

   c. internal failure costs.

   d. external failure costs.

34. The acronym COPQ represents:

   a. cost of perfect quality.

   b. cost of product quality.

    c. cost of present quality.

    d. cost of poor quality.

35. There are several reasons to consider supply base optimization. Which of the following reasons would usually not be considered?

    a. Cost reduction

    b. Risk reduction

    c. Increased quantity

    d. Increased quality

36. In five years, $5000 will be available. What is the net present value (NPV) of that money, assuming an annual interest rate of 10%?

    a. $500.00

    b. $1581.14

    c. $3104.61

    d. $5000.00

37. The net present value (NPV) costs to conduct a project are estimated to be $100,000. The NPV benefits or savings due to the project are estimated at $750,000. Compute the benefit-to-cost ratio.

    a. $0.13

    b. $7.50

    c. $100,000

    d. $750,000

38. Which perspective defines an organization's value proposition and measures how effective the organization is in creating value?

    a. Internal business processes perspective

    b. Learning and growth perspective

    c. Financial perspective

    d. Customer perspective

39. The costs associated with acceptance sampling are:

    a. appraisal costs.

    b. prevention costs.

   c. internal failure costs.

   d. external failure costs.

40. Which of the following tools would be the most appropriate for keeping management apprised of emerging technologies relating to the supply chain?

   a. SWOT analysis

   b. PEST analysis

   c. Portfolio analysis

   d. Risk analysis

41. The phrase "vital few and useful many" is applicable to the:

   a. cause-and-effect diagram.

   b. check sheet.

   c. Pareto diagram.

   d. process flow diagram.

42. Which of the following cases would be considered the least desirable in terms of the relationship of the specification to the process spread?

   a. $6\sigma < USL - LSL$

   b. $6\sigma = USL - LSL$

   c. $6\sigma > USL - LSL$

   d. $3\sigma = USL$

43. The process spread is the same as the:

   a. tolerance.

   b. quality spread.

   c. index.

   d. process variation.

44. When determining a make/buy decision, which tool would be best for analyzing historical performance?

   a. DMAIC

   b. PDCA

   c. CAPA

   d. FMEA

45. Which of the following would be the preferred method/tool for identifying requirements?

    a. Purchasing document (for example, purchase order)

    b. Supply agreement

    c. Both (a) and (b)

    d. None of the above

46. The purchasing/procurement function has two fundamental duties. These are:

    a. to select and contract with suppliers and set terms for purchased goods and services.

    b. to ensure competitive bids and select suppliers based on cost and delivery.

    c. to source suppliers at the lowest cost and seek cost reductions going forward.

    d. to select the best supplier and negotiate a fair price.

47. Supplier performance is optimized long term by:

    a. open communication.

    b. collaboration and planning.

    c. joint problem solving.

    d. all of the above.

48. _____ should be used to prioritize supplier development opportunities, escalation/exit strategy, and future sourcing preferences.

    a. Pricing and schedule fulfillment

    b. Performance metrics

    c. On-time delivery scores

    d. Willingness to reduce cost year over year

49. Purchasing/sourcing/procurement:

    a. is purely tactical.

    b. is mostly strategic.

    c. can be both strategic and tactical.

    d. is focused on cost reductions.

50. Initially, identifying and communicating roles and responsibilities with the supplier is based on:

    a. purchasing and engineering recommendations.

    b. the results of the audit.

    c. the complexity of the purchased item, and the manufacturing process required.

    d. what is needed from the supplier, based on the strength of internal core (technical) competencies.

51. The ideal supplier management relationship is based on mutual:

    a. competition.

    b. admiration.

    c. segregation.

    d. benefit.

52. The need for a supplier typically arises in the _____ phase.

    a. product development

    b. procurement

    c. production

    d. participation

53. Before suppliers provide a quotation for a new project, what should the supplier consider?

    a. Capability analysis

    b. Capacity analysis

    c. Feasibility analysis

    d. Takt time analysis

54. Once everything is agreed on between the customer and supplier, the success of the supplier deployment can be measured through:

    a. communications.

    b. monitoring.

    c. design reviews.

    d. reconciliation.

ANSWERS

1. c; A mission statement explains the company's reason for existence. A vision statement is what or where the company would like to be in the future. [I.A]

2. d; A vision statement is what or where the company would like to be in the future. The other answers are more task oriented and do not reflect the future state. [I.A]

3. d; Measurable supporting objectives are usually measurable and supported by qualifiable metrics. Purpose, mission, and vision are statements. [I.A]

4. b; While complying with regulations, delivering quality products, ensuring market share, and providing escort licenses all support supplier partnerships, the primary reason for supplier partnerships is to consistently deliver quality products. [I.A]

5. a; Commitment and trust are essential to forming partnerships. [I.A]

6. a; Suppliers–inputs–process–outputs–customers (SIPOC) can be used to identify stakeholders. [I.B.1]

7. b; Supplier life cycle management (SLM) supports and guides the business relationship from initial supplier discovery through qualification and onboarding to ongoing maintenance and possible termination or obsolescence. [I.B.1]

8. a; Supply chain management (SCM) focuses more explicitly on the entire supply chain—including lower-tier suppliers—across the entire enterprise, and focuses on managing and minimizing total systemwide costs. [I.B.1]

9. c; Suppliers–inputs–process–outputs–customers (SIPOC) can be used to identify stakeholders. [I.B.1]

10. d; SLM is an integrated approach that considers business and quality needs of the organization.

    SCM is a set of system-focused approaches utilized to efficiently integrate suppliers, manufacturers, warehouses, and stores so that merchandise is produced and distributed in the right quantities, to the right locations, and at the right time in order to minimize systemwide costs while satisfying service level requirements.

    The primary difference between SLM and SCM is that SCM focuses on supplier integration. [I.B.1]

11. b; $PI = \dfrac{\text{Purchased costs} + \text{Nonproductive costs}}{\text{Purchased costs}} = \dfrac{\$1000 + \$100}{\$1000} = 1.1.$

    [I.B.2]

12. c; Supplier C has the lowest elevated cost per part ($2018.50).

	Supplier A	Supplier B	Supplier C	Supplier D
Part cost	$1750	$1650	$1835	$1800
Performance index (PI)	1.3	1.4	1.1	1.2
	2275	2310	2018.50	2160

[I.B.2]

13. b; Supplier B has the highest elevated cost per part ($2310).

	Supplier A	Supplier B	Supplier C	Supplier D
Part cost	$1750	$1650	$1835	$1800
Performance index (PI)	1.3	1.4	1.1	1.2
	2275	2310	2018.50	2160

[I.B.2]

14. c; The elevated cost for supplier A is $2275.

	Supplier A	Supplier B	Supplier C	Supplier D
Part cost	$1750	$1650	$1835	$1800
Performance index (PI)	1.3	1.4	1.1	1.2
	2275	2310	2018.50	2160

[I.B.2]

15. d; The elevated cost for supplier B is $2310.

	Supplier A	Supplier B	Supplier C	Supplier D
Part cost	$1750	$1650	$1835	$1800
Performance index (PI)	1.3	1.4	1.1	1.2
	2275	2310	2018.50	2160

[I.B.2]

16. d; Strategic products (high profit impact, high supply risk). These products deserve the most attention from purchasing managers. Options include developing long-term supply relationships, regularly analyzing and managing risks, planning for contingencies, and producing the item in-house rather than buying it, if appropriate. [I.B.3]

17. c; Leverage products (high profit impact, low supply risk). Purchasing approaches to consider here include using your full purchasing power, substituting products or suppliers, and placing high-volume orders. [I.B.3]

18. b; Bottleneck products (low profit impact, high supply risk). Useful approaches here include over-ordering when the item is available (lack of reliable availability is one of the most common reasons that supply is disrupted) and looking for ways to control vendors. [I.B.3]

19. a; Routine products (low profit impact, low supply risk). Purchasing approaches for these items include using standardized products, monitoring and/or optimizing order volume, and optimizing inventory levels. [I.B.3]

20. d; The Kraljic portfolio segmentation model classifies products with low profit impact, high supply risk, and high sourcing difficulty as bottleneck products. [I.B.3]

21. c; Strategic products (high profit impact, high supply risk). These products deserve the most attention from purchasing managers. Options include developing long-term supply relationships, regularly analyzing and managing risks, planning for contingencies, and producing the item in-house rather than buying it, if appropriate. [I.B.3]

22. b; Routine products (low profit impact, low supply risk). Purchasing approaches for these items include using standardized products, monitoring and/or optimizing order volume, and optimizing inventory levels. [I.B.3]

23. d; Leverage products (high profit impact, low supply risk). Purchasing approaches to consider here include using your full purchasing power, substituting products or suppliers, and placing high-volume orders. [I.B.3]

24. a; Bottleneck products (low profit impact, high supply risk). Useful approaches here include over-ordering when the item is available (lack of reliable availability is one of the most common reasons that supply is disrupted) and looking for ways to control vendors. [I.B.3]

25. d; SIPOC stands for suppliers, inputs, process, outputs, and customers. The SIPOC tool is very effective in helping understand the current process so that improvements can be made. [I.B.4]

26. d; During the *act* stage project standardization would be applied to all suppliers. [I.B.4]

27. b; During the *do* stage supply chain management program development and procurement activities should be started with a few selected suppliers. [I.B.4]

28. c; During the *check* stage various tools such as benchmarking and supplier scorecards are utilized. [I.B.4]

29. a; During the *plan* stage a spend analysis and diagnosis of the value being purchased are evaluated. [I.B.4]

30. d; External failure costs are costs incurred when a failure occurs while the customer owns the product. [I.C.1]

31. a; Appraisal costs are costs associated with the inspection and appraisal processes. [I.C.1]

32. c; Internal failure costs are costs incurred when a failure occurs in-house, and are usually associated with the cost of scrap and rework. [I.C.1]

33. b; Prevention costs are the costs of all activities whose purpose is to prevent failures, including training, quality planning, and quality control activities. [I.C.1]

34. d; COPQ is the acronym used for the cost of poor quality. COPQ is categorized as internal, external, prevention, and appraisal. [I.C.1]

35. c; The objectives of supply base optimization are cost reduction, risk reduction, and increased quality. [I.C.2]

36. c; In five years, $5000 will be available. The net present value of that money, assuming an annual interest rate of 10% can be calculated by

$$P = F(1 + i)^{-n}$$

where

P = Net present value

F = Amount to be received n years from now

i = Annual interest rate expressed as a decimal

$$P = \$5000.00(1 + 0.10)^{-5} = \$3104.62$$

[I.C.2]

37. b; The net present value (NPV) costs to conduct a project are estimated to be $100,000. The NPV benefits or savings due to the project are estimated at $750,000. The benefit-to-cost ratio can be calculated by

$$\frac{\Sigma \text{ NPV of all benefits anticipated}}{\Sigma \text{ NPV of all costs anticipated}} = \frac{\$750,000}{\$100,000} = \$7.50$$

[I.C.2]

38. d; The customer perspective defines an organization's value proposition and measures how effective the organization is in creating value for its customers through its goals, objectives, strategies, and processes. [I.C.2]

39. a; Appraisal costs are costs associated with the inspection and appraisal processes. [I.C.2]

40. a; A SWOT (strengths, weaknesses, opportunities, and threats) analysis is an effective strategic planning tool applicable to a business or project objective. Strengths and weaknesses are identified with respect to the internal capabilities of an organization, while opportunities and threats look outside the organization to identify opportunities for the organization and threats to the organization. [I.C.3]

41. c; The purpose of the Pareto chart is to separate the "vital few" causes from the "trivial many." This is often reflected in what is called the 80/20 rule and helps focus attention on the more pressing issues. The 80/20 rule means that in any situation, 20 percent of the inputs or activities are responsible for 80 percent of the outcomes or results. [I.C.3]

42. a; The least desirable relationship between the specification and the process spread is when $6\sigma < USL - LSL$, as demonstrated in the figure below.

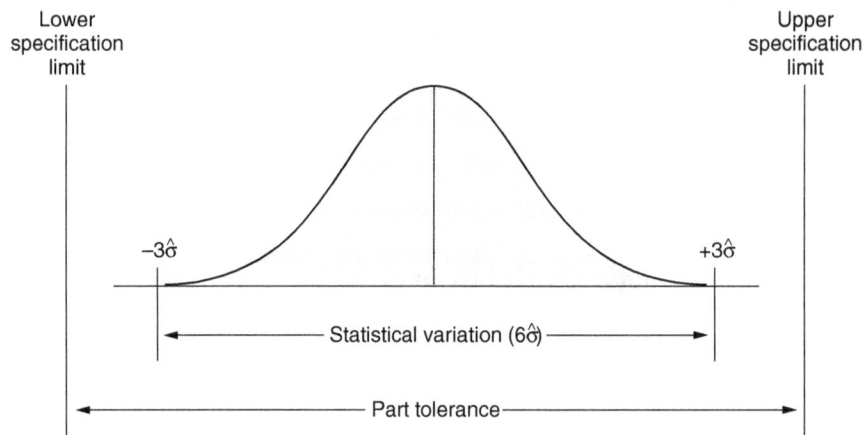

Relationship between statistical control limits and product specifications.

Source: Adapted from Durivage (2014). Used with permission.

[I.C.3]

43. d; The process spread is simply the amount of variation present in the process. [I.C.3]

44. c; Key data sources for analyzing historical performance include supplier corrective action requests (SCARs), corrective and preventive action (CAPA),

internal nonconformances, historical capability, and results from regulatory, ISO, and customer inspections and audits. Planned or scheduled maintenance, age of equipment, historical downtime, first-pass yield, and rolled throughput yield are also great predictors of success. [I.C.3]

45. c; Requirements can be found on both POs and supply agreements. It's important to recognize that terms and conditions can be found on a PO, sometimes on the back or sometimes on a website, but either way, they are binding. [I.D]

46. a; Although all four answers have more or less merit, answer (a) describes the basic purchasing function most accurately as it exists in most organizations. [I.D]

47. d; All of these activities are critical to maintaining a healthy level of supplier performance. [I.D]

48. b; (a) and (c) are important but usually a subset of performance, and without a full picture of performance, price reductions can be very dangerous and increase risk. [I.D]

49. c; Sourcing or purchasing can be both tactical and strategic. For example, expediting POs in support of production is tactical. Setting up long-term partnerships is strategic. [I.D]

50. d; (a) and (b) are not accurate answers. (c) is a subset of (d)—starting to get to the root of internal competencies. (d) is most complete. [I.E]

51. d; The ideal supplier management relationship is based on mutual benefit. However, trust and cooperation are also important to the relationship. [I.E]

52. a; The need for a supplier typically arises in the product development phase. The earlier the need for a supplier is determined, the easier a supplier can be selected and integrated into the design, development, and production phases. [I.E]

53. c; Before suppliers provide a quotation for a new project, the supplier should consider performing a feasibility analysis. Capability analysis, capacity analysis, and takt time analysis are elements of a feasibility analysis. [I.E]

54. b; Once everything is agreed on, the success of the supplier deployment can be measured through monitoring of agreed-on performance metrics. [I.E]

Part II

Risk Management

(37 questions)

A. STRATEGY

1. *System.* Develop a risk-based approach to manage the supply base, including business continuity and contingency planning. (Create)

2. *Product/service.* Develop and implement a risk mitigation plan to minimize, monitor, and/or control risks. (Evaluate)

3. *Prevention strategies.* Identify and evaluate strategies and techniques such as supply chain mapping, avoidance, detection and mitigation used to prevent the introduction of counterfeit parts, materials, and services. (Evaluate)

B. ANALYSIS AND MITIGATION

1. *Analysis.* Identify, assess and prioritize risks to supplier quality using tools such as, decision analysis (DA), failure mode effects analysis (FMEA), fault tree analysis (FTA), and process auditing. (Evaluate).

2. *Mitigation control.* Develop and deploy controls such as inspection or test plan. Prioritize mitigation activities and sustain a risk mitigation plan appropriate to the risk of the product/service. (Create)

3. *Mitigation effectiveness.* Verify the effectiveness of the control plan and improve if necessary, using continuous improvement methods such as plan–do–check–act (PDCA), lean and product auditing tools. (Create)

QUESTIONS

1. Which of the following tools would be the most appropriate for crisis management?

    a. PEST analysis

    b. SWOT analysis

    c. Contingency planning

    d. Risk analysis

2. Who is the most responsible for risk?

    a. Supplier

    b. Customer

    c. Consumer

    d. End user

3. A supplier quality engineer is participating with the development team, which has been charged with the task of redesigning a product and reducing the overall risk to the end user. Which of the following tools would be the most appropriate for the team to use?

    a. PDCA

    b. DOE

    c. FMEA

    d. SWOT

4. When an organization is determining possible risks, they should:

    a. include all possible risks, using many risk identification tools.

    b. include only those risks that are obvious and currently known.

    c. only come up with a list of no more than three to five risks or hazards.

    d. include only risks that they have control over.

5. Residual risk is:

    a. risk that exists before a risk treatment has been implemented.

    b. risk due to a gap between what is thought to be risk and what really is risk.

    c. unknown risk that can never be identified using FMEA.

    d. risk that remains after a risk treatment has been implemented.

6. The hierarchy of risk control should follow which of the following sequences?

   a. Substitute, eliminate, engineering controls, administrative controls

   b. Substitute, eliminate, administrative controls, engineering controls

   c. Eliminate, substitute, engineering controls, administrative controls

   d. Eliminate, substitute, administrative controls, engineering controls

7. The purpose of risk evaluation is to:

   a. determine options for modifying risks.

   b. compare risk level to stated risk criteria.

   c. decide whether risk levels are acceptable.

   d. evaluate and possibly change the consequences.

8. Which of the following would be part of risk monitoring and review?

   a. Identifying new or upcoming risks

   b. Obtaining new information to update risk levels

   c. Assessing the controls in place to make sure they are working and appropriate

   d. All of the above

9. Once the product or service has been defined and the suppliers have been selected, it is important that the manufacturer create a risk mitigation plan to minimize, monitor, and/or _____ risks.

   a. control

   b. conceive

   c. correlate

   d. connect

10. Postproduction risk management can be derived from various sources, including service personnel, training personnel, incident reports, and _____.

   a. production monitoring

   b. process validation

   c. customer feedback

   d. incoming inspection

11. Quality risk management supports a(n) _____ approach to decision making.

    a. anecdotal

    b. biased

    c. procedural

    d. scientific

12. Risk assessment may be undertaken in varying degrees of depth and detail using one or several methods ranging from simple to complex. Risk assessments should consider:

    a. training, education, and experience.

    b. procedures, complexity, and training.

    c. uncertainty, resources, and complexity.

    d. residuals, resources, and training.

13. Risk assessment can be applied at all stages of _____, and is usually applied many times with different levels of detail to assist in the decisions that need to be made at each phase.

    a. product realization

    b. the life cycle

    c. service and installation

    d. product design

14. Increased globalization and outsourcing has seen the introduction of _____ materials and parts within the supply chain.

    a. counterfeit

    b. unique

    c. complex

    d. complete

15. Available data do not always provide a reliable basis for predicting the future. For unique types of risks or for new product types, _____ data may not be available.

    a. attribute

b. variables

c. neutral

d. historical

16. A process improvement team has completed a PFMEA. Which of the following tools would be best suited to help the team focus their efforts based on the PFMEA?

    a. SWOT analysis

    b. Pareto analysis

    c. Force-field analysis

    d. Statistical analysis

17. A value indicating the relative risk of a potential failure is referred to as:

    a. failure.

    b. modality.

    c. RPN.

    d. severity.

18. An FMEA has a severity of 7, a probability of occurrence of 5, and a probability of detection of 3. What is the RPN?

    a. 12

    b. 15

    c. 35

    d. 105

19. A systematic approach that proactively identifies, analyzes, prioritizes, and documents potential failure modes and their respective potential causes of failures is:

    a. DFSS.

    b. FMEA.

    c. SIPOC.

    d. PDCA.

20. The likelihood that current controls will prevent a failure from reaching the customer is called:

   a. detection.

   b. occurrence.

   c. severity.

   d. FMEA.

Use the following diagram to answer questions 21–24.

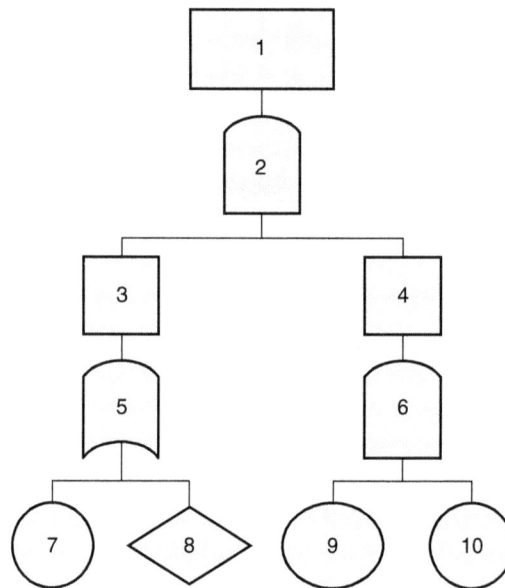

21. Which of the fault tree analysis (FTA) symbols indicates an AND gate?

   a. 2

   b. 5

   c. 3

   d. 2 and 6

22. Which of the fault tree analysis (FTA) symbols indicates an OR gate?

   a. 2

   b. 5

   c. 3

   d. 6

23. Which of the fault tree analysis (FTA) symbols indicates an undeveloped event?

    a. 7

    b. 8

    c. 9

    d. 7 and 10

24. Which of the fault tree analysis (FTA) symbols indicates the primary event?

    a. 1

    b. 3

    c. 5

    d. 3 and 4

25. _____ needs to be an integral part of the risk treatment plan to ensure that the measures remain effective.

    a. Mitigation

    b. Reporting

    c. Monitoring

    d. Planning

26. An effective supplier management program can reduce the risk involved when there are routine _____ with the supplier.

    a. communications

    b. mitigation controls

    c. purchase orders

    d. inspections

27. Labeling, warning, identification, traceability, risk management, and recall requirements are all part of an effective:

    a. product marketing campaign.

    b. supplier quality agreement.

    c. compliance program.

    d. risk mitigation program.

28. Sampling plans provide for effective and economical inspection. Although economical, sampling does provide a degree of risk to the:

    a. producer.

    b. consumer.

    c. regulators.

    d. producer and consumer.

29. The type of risk involved when a lot could be rejected when it does conform to the specification is known as:

    a. Ω.

    b. α.

    c. β.

    d. σ.

30. The type of risk involved when a lot could be accepted when it does not conform to the specification is known as:

    a. Ω.

    b. α.

    c. β.

    d. σ.

31. Analyzing the problem is associated with which phase of the PDCA cycle?

    a. Plan

    b. Do

    c. Check

    d. Act

32. Studying the results is associated with which phase of the PDCA cycle?

    a. Plan

    b. Do

    c. Check

    d. Act

33. Standardizing the solution is associated with which phase of the PDCA cycle?

    a. Plan

    b. Do

    c. Check

    d. Act

34. Identifying the opportunity is associated with which phase of the PDCA cycle?

    a. Plan

    b. Do

    c. Check

    d. Act

35. Implementing the solution is associated with which phase of the PDCA cycle?

    a. Plan

    b. Do

    c. Check

    d. Act

36. Developing the optimal solution is associated with which phase of the PDCA cycle?

    a. Plan

    b. Do

    c. Check

    d. Act

37. Standardizing the solution is associated with which phase of the PDCA cycle?

    a. Plan

    b. Do

    c. Check

    d. Act

ANSWERS

1. c; Contingency planning (also called a *plan B*) is used for crisis management, business continuity, and asset security. [II.A.1]

2. b; Although a supplier may conduct risk management activities, the customer is ultimately responsible for risk. [II.A.1]

3. c; The most appropriate tool for a team charged with the task of redesigning a product and reducing the overall risk to the end user would be FMEA. [II.A.1]

4. a; The purpose of risk identification is to identify all possible risks, whether current, possible future risks, risks that are not currently under the organization's control, or risks that may occur due to the results of an accumulation of factors or steps in the process. [II.A.1]

5. d; Residual risk is that risk remaining even after a treatment of a known risk has been implemented. [II.A.1]

6. c; The hierarchy of risk control should eliminate, substitute, provide engineering controls, and administrative controls. [II.A.2]

7. b; Risk evaluation is used to make decisions about which risks need to be addressed. [II.A.2]

8. d; Risk monitoring and review consists of identifying new or upcoming risks, obtaining new information to update risk levels, and assessing the controls in place to make sure they are working and appropriate. [II.A.2]

9. a; Once the product or service has been defined and the suppliers have been selected, it is important that the manufacturer create a risk mitigation plan to minimize, monitor, and/or control risks. [II.A.2]

10. c; Postproduction information can be part of established QMS procedures (for example, monitoring and measuring). Manufacturers must establish procedures to collect information from various sources, such as service personnel, training personnel, incident reports, and customer feedback. The other choices are related to preproduction or production activities. [II.A.2]

11. d; Quality risk management supports a scientific approach to decision making. [II.A.3]

12. c; Risk assessment may be undertaken in varying degrees of depth and detail using one or several methods ranging from simple to complex. Risk assessments should consider degree of uncertainty, availability of resources, and complexity. [II.A.3]

13. b; Risk assessment can be applied at all stages of the life cycle, and is usually applied many times with different levels of detail to assist in the decisions that need to be made at each phase. Life cycle phases have different needs and require different techniques. [II.A.3]

14. a; Increased globalization and outsourcing has seen the introduction of counterfeit materials and parts within the supply chain. Companies that sell counterfeit materials and parts because of loose supply chain oversight need to face serious penalties, and cannot excuse themselves by blaming their suppliers. [II.A.3]

15. d; Available data do not always provide a reliable basis for predicting the future. For unique types of risks or for new product types, historical data may not be available or there may be different interpretations of available data by different stakeholders (industry and regulators, for example). [II.A.3]

16. b; A Pareto chart can be used to help a process improvement team focus their efforts based on the PFMEA. [II.B.1]

17. c; The RPN 24 is a value indicating the relative risk of the potential failure. The RPN is the product of the severity, probability of occurrence, and probability of detection. [II.B.1]

18. d; To calculate the RPN for an FMEA with a severity of 7, a probability of occurrence of 5, and a probability of detection of 3, the following formula is used:

$$RPN = S \times O \times D = 7 \times 5 \times 3 = 105$$

where

S = Severity

O = Probability of occurrence

D = Probability of detection

[II.B.1]

19. b; FMEA is a systematic approach that proactively identifies, analyzes, prioritizes, and documents potential failure modes and their respective potential causes of failures. [II.B.1]

20. a; Detection is the likelihood that current controls will prevent a failure from reaching the customer. [II.B.1]

21. d; Numbers 2 and 6 represent AND gates.

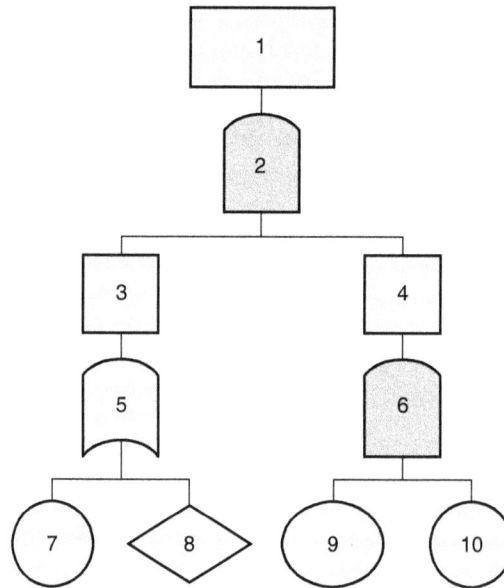

[II.B.1]

22. b; Number 5 represents an OR gate.

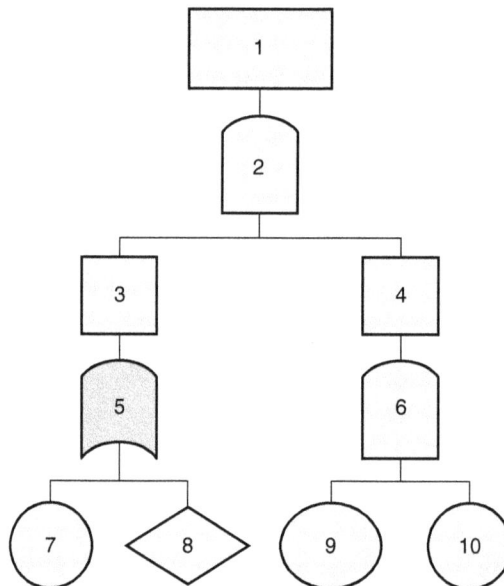

[II.B.1]

23. b; Number 1 represents an undeveloped event.

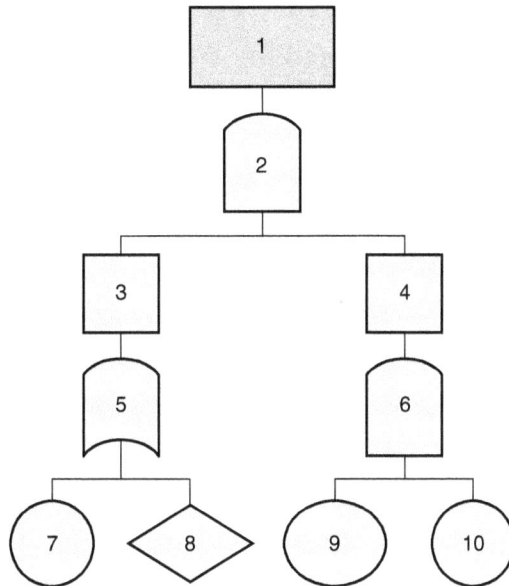

[II.B.1]

24. a; Number 1 represents a primary event.

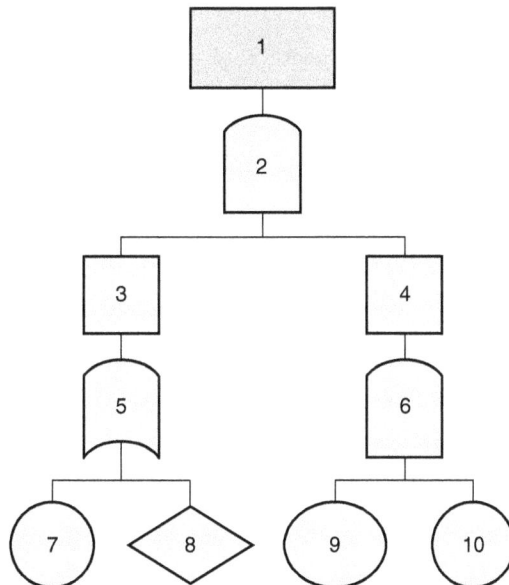

[II.B.1]

25. c; Monitoring needs to be an integral part of the risk treatment plan to ensure that the measures remain effective. [II.B.2]

26. a; An effective supplier management program can reduce the risk involved when there are routine communications with the supplier. [II.B.2]

27. d; Labeling, warning, identification, traceability, risk management, and recall requirements are all part of an effective risk mitigation program. [II.B.2]

28. d; Sampling plans provide for effective and economical inspection. Although economical, sampling does provide a degree of risk to the producer and consumer. [II.B.2]

29. b; The type of risk involved when a lot could be rejected when it does conform to the specification is known as *alpha (α) risk*. [II.B.2]

30. c; The type of risk involved when a lot could be accepted when it does not conform to the specification is known as *beta (β) risk*. [II.B.2]

31. a; Analyzing the problem is associated with the *plan* phase of the PDCA cycle. [II.B.3]

32. c; Studying the results is associated with the *check* phase of the PDCA cycle. [II.B.3]

33. d; Standardizing the solution is associated with the *act* phase of the PDCA cycle. [II.B.3]

34. a; Identifying the opportunity is associated with the *plan* phase of the PDCA cycle. [II.B.3]

35. b; Implementing the solution is associated with the *do* phase of the PDCA cycle. [II.B.3]

36. a; Developing the optimal solution is associated with the *plan* phase of the PDCA cycle. [II.B.3]

37. d; Standardizing the solution is associated with the *act* phase of the PDCA cycle. [II.B.3]

Part III

Supplier Selection and Part Qualification

(60 questions)

A. PRODUCT/SERVICE REQUIREMENTS DEFINITION

1. *Internal design reviews.* Identify and apply common elements of the design review process, including roles and responsibilities of the participants. (Apply)

2. *Identifying requirements.* Identify and apply internal requirements (e.g. interrelated functional business units) for product or service in collaboration with stakeholders, including the requirements for supply chain and sub-tier suppliers. (Evaluate)

B. SUPPLIER SELECTION PLANNING

1. *Supplier comparison.* Evaluate existing supplier's capabilities, capacities, past quality, delivery, price, lead times, and responsiveness against identified requirements. (Evaluate)

2. *Potential suppliers evaluation.* Assess potential new suppliers against identified requirements using tools such as, self-assessments, audits and financial analysis. Verify third-party certification status and regulatory compliance and analyze and report on results of assessments to support the supplier selection process. (Evaluate)

3. *Supplier selection.* Evaluate and select supplier based on analysis of assessment reports and existing supplier evaluations, using decision analysis tools and selection matrices. (Evaluate)

C. PART, PROCESS AND SERVICE QUALIFICATION

1. *Technical review.* Interpret and evaluate technical specification requirements and characteristics such as, views, title blocks, dimensioning and tolerancing and GD&T symbols as they relate to product and process. (Evaluate)

2. *Supplier relations.* Collaborate with suppliers to define, interpret, and classify quality characteristics for the part/process/service. (Evaluate)

3. *Process and service qualification planning.* Develop a part/process/service qualification plan with supplier and internal team, that includes calibration requirements, sample size, first article inspection, measurement system analysis (MSA), process flow diagram (PFD), failure mode and effects analysis (FMEA), control plans, critical to quality (CTQ), inspection planning, capability studies, material and performance testing, appearance approval and internal process validation. (Analyze)

4. *Part approval.* Understand the production part approval process (PPAP) requirements and ensure suppliers understand the processes required to produce parts with consistent quality during an actual production run at production rates. (Understand)

5. *Validate requirements.* Collaborate with internal team to interpret the results of the executed qualification plan for the part/process/service. (Evaluate)

QUESTIONS

1. The _____ shows the relationship between requirements, risks, and needs of the customer.

    a. demand planning

    b. house of quality

    c. PFMEA

    d. design output

2. The APQP process is designed to be used during:

    a. the product development process.

    b. the RFQ process.

    c. the supplier qualification process.

    d. the product validation process.

3. _____ define(s) the phases of design, frequency of design reviews, controls on the product development process, and review and approval authorities.

    a. Design and development planning

    b. Design and development outputs

    c. Design and development changes

    d. Design and development inputs

4. _____ must be reviewed for adverse effects, approved, and controlled throughout the process.

    a. Design and development planning

    b. Design and development outputs

    c. Design and development changes

    d. Design and development inputs

5. The organization must ensure that _____ satisfy/satisfies the inputs requirements.

    a. design and development planning

    b. design and development outputs

    c. design and development changes

    d. design and development inputs

6. During design and development, _____ must consider functional and performance requirements, regulatory requirements, and safety requirements.

   a. design and development planning

   b. design and development outputs

   c. design and development changes

   d. design and development inputs

7. An often overlooked component of supplier qualification is:

   a. the audit.

   b. sub-tier suppliers.

   c. financial due diligence.

   d. the supplier questionnaire.

8. Requirements will vary based on:

   a. cost of goods sold.

   b. the capability of the supplier to provide the purchased goods or services.

   c. the nature and complexity of the purchased goods or services.

   d. the piece part price negotiated by procurement.

9. _____ is a balancing process best done by internal and external stakeholders in an atmosphere of collaboration.

   a. Finalizing cost

   b. Finalizing requirements

   c. Finalizing quality agreements

   d. Finalizing preparations for an audit

10. The objective of finalizing requirements collaboratively is:

    a. to simultaneously meet goals for quality, cost, and delivery.

    b. to make sure all parties agree.

    c. so the supplier can't say later "I didn't know."

    d. to make sure purchasing meets their spend reduction targets.

11. Internal stakeholders for requirements include:

    a. R&D/engineering.

    b. procurement/purchasing.

    c. operations.

    d. all of the above.

12. Procurement/purchasing function controls are applied to:

    a. prevent purchasing from going with the lowest price only.

    b. ensure the quality of the organization's procurement function.

    c. ensure outsourced product/service quality.

    d. (a) and (c).

13. The process for current supplier comparison should be defined because:

    a. predefined processes ensure fairness and accuracy in comparison.

    b. predefined processes ensure consistency in application, avoiding the appearance of any impropriety.

    c. predefined processes are relational.

    d. both (a) and (b).

14. Capabilities, price, capacity, quality history, and responsiveness are:

    a. great categories for a scorecard.

    b. great categories for supplier comparison.

    c. nice-to-have categories after price and delivery.

    d. great questions to ask on a desktop audit.

15. The "best ranking approach" uses what type of tool for analysis?

    a. Return-on-investment analysis

    b. Cost–benefit analysis

    c. A table to tabulate results

    d. A decision matrix

For questions 16–18, refer to the following table. Higher values are more desirable.

Supplier	Price	Weight	Quality	Weight	Capacity	Weight	Totals
A	9	1	6	1	3	2	21
B	3	1	6	1	9	2	27
C	6	1	9	1	3	2	21
D	6	1	9	1	6	2	27

16. Based on the table, what's the most important category for comparison?

    a. Price

    b. Quality

    c. Capacity

    d. All are equal

17. Which supplier is most desirable?

    a. A

    b. B

    c. C

    d. D

18. Which approach does this table represent?

    a. FMEA RPN

    b. Control plan

    c. The "best ranking approach"

    d. Risk–benefit analysis

19. A potential supplier's quality management system (QMS) exhibits continuous monitoring and updating for necessary changes and emerging leading practices. The QMS would be considered:

    a. defined.

    b. managed.

    c. repeatable.

    d. optimized.

20. Handling the supplier qualification audit well, including open communication, collaboration, and meeting commitments, is important because:

    a. compliance is a critical requirement for suppliers to establish and maintain.

    b. it might well be a "first impression" for both parties and therefore critical to the future relationship.

    c. it is a necessity to begin the process of onboarding suppliers.

    d. it can give insights into the supplier's systems and controls that a questionnaire can't.

21. A financial audit is critical to:

    a. establish long-term economic viability and stability.

    b. establish compliance with applicable policies and regulations.

    c. quality certification.

    d. both (a) and (b).

Choose from the following list to answer questions 22–24.

    A. A step-by-step review of inputs and outputs that relate to a specific operation, or combination of operations.

    B. A review of the implementation regulatory requirements, and adherence.

    C. A review of elements governing processes, policies, and records for the assurance of quality.

    D. A review of critical-to-quality characteristics to understand performance measures such as out-of-the-box failure.

22. A process audit is:

    a. A.

    b. B.

    c. C.

    d. D.

23. A QMS audit is:

    a. A.

    b. B.

    c. C.

    d. D.

Part III
Questions

24. A product audit is:

    a. A.

    b. B.

    c. C.

    d. D.

25. Supplier selection activities can all be boiled down to _____ to choose the best overall supplier.

    a. characterizing comparative risks of several factors

    b. determining the lowest cost and highest quality

    c. understanding the supplier's comparative capacity and lowest loss time

    d. determining the supplier with the highest comparative capability and (potential) service levels

26. Documentation of supplier selection activities must be maintained to ensure _____ over time.

    a. record control

    b. the integrity of the supplier file

    c. consistency

    d. fairness

Use the following risk table to answer questions 27–30. Higher values are less desirable.

Supplier	QMS maturity	Weight	Value	Financial	Weight	Value	Capacity	Weight	Value	Totals
A	2	2	4	3	1	3	3	3	9	108
B	2	2	4	4	1	4	2	3	6	96
C	1	2	2	2	1	2	2	3	6	24
D	4	2	8	1	1	1	3	3	9	72

27. _____ is the most critical category.

    a. QMS maturity

    b. Capacity

    c. Finance

    d. All are equal

28. Which is the lowest-risk supplier?

    a. A

    b. B

    c. C

    d. D

29. Which is the highest-risk supplier?

    a. A

    b. B

    c. C

    d. D

30. Which supplier has the least quality risk?

    a. A and B are tied

    b. C

    c. D

    d. A only

31. An engineer needs to copy an ISO A2 drawing. What sheet size (in millimeters) does the engineer need?

    a. 841×1189

    b. 594×841

    c. 420×594

    d. 297×420

32. Drawing dimensions can be linear, circular, or:

    a. rectangular.

    b. triangular.

    c. angular.

    d. singular.

33. The technical review:

    a. establishes that requirements are fully understood and accepted by suppliers.

    b. highlights technical capability of the supplier.

    c. allows for supplier input.

    d. is meaningless if the supplier objects to some or all requirements.

34. Other requirements in the technical review can include:

    a. manufacturing environment (for example, cleanroom).

    b. functional specifications.

    c. material properties.

    d. all of the above.

35. Points of discussion with the supplier might also include:

    a. how their process can affect the finished product.

    b. future cost reductions.

    c. run rate.

    d. (a) and (c).

36. It is a best practice when concluding a technical review to:

    a. thank the supplier.

    b. review the items discussed.

    c. document the agreed-on output of the meeting.

    d. rely on memory.

37. Why is it important to *collaborate* with the supplier and get supplier feedback on requirements?

    a. It's not. Suppliers should listen to their customers and follow the contract.

    b. The supplier needs to own adherence to the specs once they are agreed to.

    c. The supplier is an expert in the process they perform and has a great deal of experience translating customer requirements into their processes.

    d. The supplier most likely has more process knowledge about their process.

Part III
Questions

38. One very important reason to collaborate with the supplier on critical requirements is that:

    a. sometimes process control characteristics are more important to monitor than are design-critical characteristics, and the depth of knowledge at the supplier can help determine that.

    b. it's important to have a mutual agreement on quality plans, inspection, tolerances, and other requirements such as environmental controls.

    c. it helps to establish and/or deepen the relationship with the supplier, as well as obtain their buy-in.

    d. while collaboration is great, design requirements are nonnegotiable, so the relationship might need to be leveraged.

39. Collaboration on requirements also adds value in that:

    a. it builds the relationship, providing a solid foundation for later in the project.

    b. it can provide the supplier with knowledge of how the customer will use their product.

    c. it provides a healthy give-and-take on both parties' behalf.

    d. it helps the supplier feel ownership and encourages buy-in.

40. When using GD&T, MMC refers to:

    a. median material condition

    b. mean material condition

    c. minimum material condition

    d. maximum material condition

Use the following figure to answer question 41.

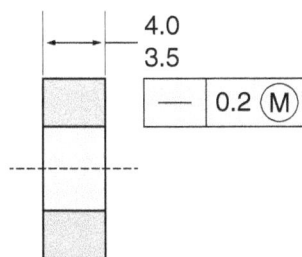

41. What is the bonus tolerance if the feature is 3.7?

    a. 0.0

    b. 0.2

    c. 0.3

    d. 0.5

Use the following figure to answer question 42.

$$2 \times \varnothing \quad \begin{matrix} 4.0 \\ 3.5 \end{matrix}$$

$$\boxed{\oplus \;|\; \varnothing\; 0.1\; \text{(M)}}$$

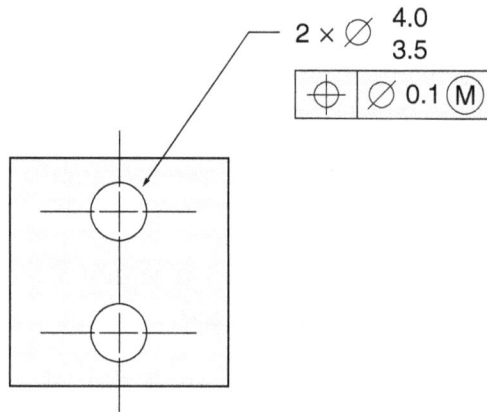

42. What is the bonus tolerance if the feature is 3.7?

    a. 0.0

    b. 0.2

    c. 0.3

    d. 0.5

43. A process flow diagram (PFD) does which of the following?

    a. Controls the process

    b. Diagrams a lean material system using the principle of flow

    c. Maps the process steps and sequence of operations in a process

    d. Displays where product goes in the value stream

44. A PFMEA is a _____ tool.

    a. risk management

    b. quality management

    c. sales forecasting

    d. quality engineering

45. What is the primary relationship between the process flow diagram (PFD) and the control plan?

    a. The control plan must be created first, as it plans the control process.

    b. There is no relationship; they are independent.

    c. They are both components of PPAP.

    d. The control plan addresses each operating step laid out in the PFD.

46. The production part approval process (PPAP):

    a. has no value outside of the automotive industry.

    b. establishes that components and subassemblies meet requirements.

    c. is too difficult for most suppliers.

    d. includes the elements of quoting, costing, and demand planning.

47. Gage R&R studies are important because:

    a. they establish measurement system error and, if acceptable, ensure that the data we collect for other acceptance activities are accurate and precise.

    b. these studies prove through objective evidence whether an inspector is proficient.

    c. these studies qualify inspectors to complete PPAP.

    d. these studies highlight the differences in technique between inspectors.

48. CTQ characteristics:

    a. are critical to quality.

    b. highlight technical capability of the supplier.

    c. allow for supplier input.

    d. both (a) and (c).

49. The technical review:

    a. is not part of PPAP.

    b. highlights technical capability to the supplier.

    c. allows for supplier input.

    d. is a one-way conversation.

50. PPAP is required in all of the following situations *except*:

    a. new parts.

    b. pricing changes.

    c. changes in part processing.

    d. sub-tier supplier or materials change.

51. The specific phrase, "establishing through objective evidence that a process consistently produces a result or product meeting its predetermined requirements" refers to:

    a. process validation.

    b. a qualified process.

    c. PPAP.

    d. process capability.

52. Process validation consists of:

    a. running at rate, RFQ, process optimization.

    b. control plans, PSW, process map.

    c. material performance tests, biocompatibility, and sterilization.

    d. IQ, OQ, and PQ.

53. In PPAP, FAI stands for:

    a. final acceptance inspector.

    b. factory acceptance initiation.

    c. first article inspection.

    d. failure and inspection.

54. A *critical-to-quality* (CTQ) characteristic is usually determined by:

    a. the effect on form/fit/function.

    b. customer perception of the overall quality of that characteristic.

    c. the need to inspect this characteristic.

    d. (a) and (c).

55. As a rule, the internal (customer) cross-functional team working on part/process approval should consist of:

    a. supplier, purchasing, and QA.

    b. QA, engineering, and operations.

    c. sales, purchasing, and operations.

    d. supplier, QA, and purchasing.

56. Collaboration with the internal team becomes especially important when:

    a. deviations from the qualification plan occur.

    b. the supplier raises prices.

    c. the supplier produces a quality part but at less than the rate required to meet demand.

    d. (a) and (c).

57. Deviations from specifications might be acceptable if:

    a. engineering signs off on *use as-is* status.

    b. procurement gets the supplier to lower their piece part price sufficiently to pay for more incoming inspection.

    c. there is sufficient engineering/scientific objective evidence there will be no adverse effect on form/fit/function, performance, or safety.

    d. there is sufficient need to launch the product on time and within budget.

58. Critical elements of PPAP to review and approve are:

    a. control plan, MSA, PFMEA, PFD, capability study.

    b. IQ, OQ, PQ.

    c. PPV, run at rate, and tooling cost.

    d. CTQs, safety program, employees training program.

59. Capability studies (assuming using C_{pk}) show all of the following except:

    a. how many parts are predicted to be out of spec, based on the C_{pk} result.

    b. how centered the distribution is within spec limits.

    c. whether parts are functional and mate with other components well.

    d. the relationship of the process spread to the specification limits.

60. When reviewing capability studies, it's important that we know that:

    a. the inspectors are adequately trained, so we can rely on the results.

    b. the process is in a state of statistical control.

    c. the resulting frequency distribution is "normally distributed."

    d. both (b) and (c).

Part III
Questions

ANSWERS

1. b; The house of quality is a graphical representation of these often competing and complex drivers. Demand planning determines how many parts we need, FMEA is a risk tool, and the house of quality can be used to develop design inputs, not outputs. [III.A.1]

2. a; APQP is not related to RFQ, but validation might be a part of it. Supplier qualification might be a part of the development process. But (a) is the right answer; it is a quality process to be used during the design and development process. [III.A.1]

3. a; Planning determines the scope and deliverables of the design process. [III.A.1]

4. c; Changes must be controlled and reviewed for impact and unintended or adverse effects. [III.A.1]

5. b; Outputs must satisfy inputs. It's how we know we've succeeded. [III.A.1]

6. d; Inputs define the product and must meet a whole host of requirements as well as satisfy the design intent. [III.A.1]

7. b; Sub-tier suppliers can be critical to quality. Understand the level and nature of outsourcing by your suppliers. Financial assessment is rarely forgotten and (a) and (d) are very similar. [III.A.2]

8. c; Requirements are requirements and are based on design outputs, as determined by the nature and complexity of your purchase. Requirements should never be altered because of price or supplier capability. [III.A.2]

9. b; Finalizing requirements is the only process listed that involves all the internal and external stakeholders. It is intended to balance stakeholder needs in order to meet goals for quality, cost, and delivery. [III.A.2]

10. a; Although (b) and (c) are technically true, the real objective is to meet goals for quality, cost, and delivery; (d) is a nonanswer as it has no bearing. [III.A.2]

11. d; All three of these groups need to be considered internal stakeholders along with supplier quality and any local purchasing or materials planners. [III.A.2]

12. d; Both to ensure the procurement process and the outsourced product/service quality. [III.A.2]

13. d; (a) and (b) are both very good reasons to use a defined process to come to the best and most defensible decision. [III.B.1]

14. b; These are categories for comparison when deciding who to give additional business to. Capacity is not usually a scorecard (monitoring) category. Quality history is not usually a category on a questionnaire or desktop audit either,

as that information does not come from the supplier, but the customer (you). [III.B.1]

15. c; This approach uses a simple table with weighted values and comparison categories. [III.B.1]

16. c; Since capacity is weighted higher, it would be the most critical category. [III.B.1]

17. b; Since there is a numerical tie, and capacity is weighted higher, supplier B would be the choice. [III.B.1]

18. c; "The best ranking approach." This uses a table to compare and tabulate numerical rankings based on the most important categories. [III.B.1]

19. d; A quality management system (QMS) that exhibits continuous monitoring and updating for necessary changes and emerging leading practices would be considered to be optimized. [III.B.2]

20. b; Although each answer is a good reason to handle an audit process well, the specific examples (that is, communication and collaboration) indicate relationship management. [III.B.2]

21. d; Both compliance (especially with anticorruption laws) and financial viability are important reasons to perform due diligence. [III.B.2]

22. a; A focus on inputs, outputs, and steps is an indication that a process is being audited. [III.B.2]

23. c; A focus on quality assurance and policies is an indication that a QMS is being audited. [III.B.2]

24. d; Critical-to-quality and out-of-the-box should lead the reader to understand that a product audit is underway. [III.B.2]

25. a; "Overall" is the operative word, so it is about risk over many categories. [III.B.3]

26. c; To insure fairness, consistency is the key. Yes, these activities are records for the supplier file, but that's not the key concept. [III.B.3]

27. b; Capacity is weighted at a 3, the highest of all the categories. [III.B.3]

28. c; (c) is the lowest at 24. [III.B.3]

29. a; (a) is the highest by far; the risk value is 108. The next closest is (b) at 96. [III.B.3]

30. c; (c) is the lowest risk at 1. (a) and (b) are tied at 2. [III.B.3]

31. c; An ISO A2 drawing sheet is 420 mm × 594 mm. [III.C.1]

32. c; Drawing dimensions can be linear, circular, or angular. [III.C.1]

33. a; The technical review is critical to open dialogue with suppliers, obtain feedback and commitment to requirements, ensure supplier capability, and get a process reality check. [III.C.1]

34. d; All of these should be included. [III.C.1]

35. d; Future cost reductions are not an element of technical review. [III.C.1]

36. c; It is *always* a best practice to document results of a supplier meeting, but more so as part of technical review. [III.C.1]

37. c; Although (b) and (d) are also great answers and are true, at the evaluation level, (c) provides the true reasoning behind the mutual value of collaboration. [III.C.2]

38. a; All four answers are correct to some degree, in some context, but once again the real value of collaboration is tapping into the profound process knowledge of your supplier. [III.C.2]

39. b; (a) is a great answer, but (b) provides more value. (c) and (d) are true but not specifically value-added regarding requirements. [III.C.2]

40. d; MMC indicates maximum material condition. For example, the smallest hole or the largest diameter allowed. [III.C.2]

41. c; Bonus tolerance = Absolute difference between MMC and actual condition. BT = 4.0 − 3.7 = 0.3. [III.C.2]

42. b; Bonus tolerance = Absolute difference between MMC and actual condition. BT = 3.5 − 3.7 = .02. [III.C.2]

43. c; A PFD is a diagram of value-added steps in the process. [III.C.3]

44. a; A PFMEA is a tool characterizing risks and their mitigations following the PFD. [III.C.3]

45. d; A control plan addresses unmitigated risks from the PFMEA, as delineated step-by-step from the PFD. [III.C.3]

46. b; PPAP is a series of quality tools designed to show that requirements are met via GRR, capability studies, SPC, and PFMEA. [III.C.3]

47. a; GRR is critical in establishing the validity of future data because it determines error in the entire measurement system: gages, staging and fixtures, and operators (inspectors). [III.C.3]

48. d; CTQs are the critical-to-quality characteristics, and it is wise to allow supplier input on this determination, as process parameters can also be CTQ. [III.C.3]

49. c; Technical review is a critical component of PPAP and is intended as a two-way dialogue with the supplier. [III.C.4]

50. b; (a), (c), and (d) are all conditions where PPAP should be performed. Pricing alone is not. [III.C.4]

51. a; Process validation includes capability studies and certainly is used to qualify processes. PPAP is an automotive tool focused on part approval. Validation is a biomedical/pharmaceutical tool that focuses on the process to ensure part quality. [III.C.4]

52. d; IQ, OQ, and PQ are specific to process validation for biomedical/pharma products. [III.C.4]

53. c; FAI is the abbreviation for first article inspection, the systematic layout and 100 percent verification of the part to the engineering drawing, an anchor in PPAP. [III.C.4]

54. d; (a) and (c) are criteria for selecting CTQs. Although checking a CTQ is usually desirable, that's the aftereffect of selecting the CTQ characteristic, not the criteria. [III.C.4]

55. b; Because part and process qualification is technical, QA, engineering, and operations are the right team, at a minimum, to evaluate the supplier's work. [III.C.5]

56. d; Deviations and failure to do *run at rate* are critical and must be discussed cross-functionally. [III.C.5]

57. c; Deviations should only be acceptable if there are engineering studies demonstrating that the change will not have adverse or unintended consequences. [III.C.5]

58. a; IQ, OQ, and PQ are never PPAP elements. PPV and tooling cost are not PPAP elements, but purchasing data. CTQs and safety are great audit elements. [III.C.5]

59. c; C_{pk} will show all of the answers except (c). Fit and performance are design outputs, unrelated to process capability. [III.C.5]

60. d; Capability indices (C_{pk}, C_p) are only valid under the assumptions of control and normality. Gage R&R shows measurement system quality (including inspector repeatability). [III.C.5]

Part IV

Supplier Performance Monitoring and Improvement

(60 questions)

A. SUPPLIER PERFORMANCE MONITORING

1. *Supplier metrics.* Define, implement, and monitor supplier performance metrics such as, quality, delivery, cost and responsiveness. (Evaluate)

2. *Supplier performance.* Analyze supplier performance data (e.g. warranty analysis/field returns, defect rates) and develop periodic reports (e.g. scorecard, dashboards). (Analyze)

3. *Supplier process performance.* Apply lean principles and applications such as 5S, Kaizen, value stream mapping, single minute exchange of dies (SMED), kanban, muda, standardized work, takt time and error proofing to reduce waste and increase performance. (Evaluate)

B. ASSESS NONCONFORMING PRODUCT/PROCESS/SERVICE

Assess and evaluate nonconforming materials to determine whether a material review board (MRB) requires disposition. Conduct risk assessments to prevent future discrepancies. (Evaluate)

C. SUPPLIER CORRECTIVE AND PREVENTIVE ACTION (CAPA)

1. *Root cause analysis tools and methods.* Evaluate the root cause analysis of a problem using tools such as, cause and effect diagrams (CE), Pareto analysis, 5 Why's, fault tree analysis, design of experiments (DOE), brainstorming, check sheets, measurement system analysis (MSA), production records and review of process flow. (Evaluate)

2. *Collaboration with supplier.* Evaluate and implement corrective/preventive action, and review its effectiveness and robustness with supplier. Understand the process of updating failure mode and effects analysis (FMEA) and process control plan, understand statistical process control (SPC), product and process design change. (Evaluate)

QUESTIONS

1. Incoming acceptance can be done through either lot or part acceptance. Acme Manufacturing, a supplier of key components, shipped 27 lots in Q1, of which one was rejected. What is the percentage of lot acceptance for Q1?

    a. 3.70

    b. 3.85

    c. 96.30

    d. 100.00

2. Incoming acceptance can be done through either lot or part acceptance. Acme Manufacturing, a supplier of key components, shipped 27 lots in Q1, of which one was rejected. What is the percentage of lots rejected for Q1?

    a. 3.70

    b. 3.85

    c. 96.30

    d. 100.00

3. In-process acceptance is calculated by part acceptance rates. Acme Manufacturing, a supplier of key components, provided the following information for Q4: 768 parts were manufactured, of which 17 parts were found to be incorrect and had to be discarded. What is the in-process acceptance percentage rate for Q4?

    a. 2.21

    b. 45.18

    c. 97.79

    d. 100.00

4. Which tool would be the most useful for analyzing failure trends?

    a. Cause-and-effect charts

    b. Pareto charts

    c. Data charts

    d. Control charts

5. A small manufacturing company uses a risk-based SCAR decision matrix. To effectively use the matrix, what is the proper process flow?

   a. Determine the frequency, identify the severity-frequency rating, identify the risk severity, take appropriate action

   b. Identify the severity-frequency rating, determine the frequency, identify the risk severity, take appropriate action

   c. Identify the risk severity, identify the severity-frequency rating, determine the frequency, take appropriate action

   d. Identify the risk severity, determine the frequency, identify the severity-frequency rating, take appropriate action

6. What does the acronym SCAR mean?

   a. Single corrective action request

   b. Supplier corrective action request

   c. Single corrective action requirement

   d. Supplier corrective action requirement

7. Which tool is best used for formally communicating quality issues with a supplier?

   a. DOE

   b. DMAIC

   c. SCAR

   d. PDCA

Use the following information to answer questions 8–10.

A supplier quality engineer is completing the quarterly scorecard for a high-risk supplier using the following criteria.

Product quality:

$$\frac{\text{Total number of line items} - \text{Number of SNCRs}}{\text{Total number of line items}} \times 60$$

On-time delivery:

$$\frac{\text{Number of on-time deliveries}}{\text{Total number of line items}} \times 40$$

Scorecard rating = Product quality + On-time delivery

Total line items for Q3: 25 items

Total SCARs: 3

Number of on-time deliveries: 20

8. The product quality score is:

   a. 32.0.

   b. 40.0.

   c. 52.8.

   d. 60.0.

9. The on-time delivery score is:

   a. 32.0.

   b. 40.0.

   c. 52.8.

   d. 60.0.

10. The supplier Q3 scorecard rating is:

    a. 40.0.

    b. 60.0.

    c. 84.8.

    d. 100.0.

11. Lean manufacturing principles are the foundation of any successful manufacturing company. The implementation of these principles can be done in any order as long as the implementation is effective and sustainable. Which of the following is least likely to be used for a lean initiative?

    a. 5S

    b. Kanban

    c. Standard work

    d. Design of experiments

12. Single-minute exchange of die (SMED) generally has a target time of _____ minute(s).

    a. 1

    b. 10

    c. 30

    d. 60

13. When developing standard work, the current best practice utilizes:

    a. figures.

    b. tables.

    c. narrative.

    d. photographs.

14. Which of the following is considered by many experts to be the sixth S in a 5S program?

    a. Safety

    b. Sort

    c. Shine

    d. Sustain

15. A supplier is using a *pull* system for managing inventory. Inventory management pull systems are also known as:

    a. Hoshin

    b. Kaizen

    c. Kanban

    d. SMED

16. Costs associated with inventory are generally considered to be approximately _____ of the actual total dollar value at the component or piece part level.

    a. 10–20%

    b. 30–40%

    c. 50–60%

    d. 60–70%

17. When developing standard work, the scope of work should be outlined or documented on a:

    a. check sheet.

    b. kanban card.

    c. flowchart.

    d. control plan.

18. Procedures and work instructions require proper:

    a. credentials.

    b. formatting.

    c. pagination.

    d. revision control.

19. For general manufacturing, average inventory turns range from:

    a. one to three.

    b. three to five.

    c. five to seven.

    d. seven to nine.

20. Kanban uses the _____ to determine the right amount of inventory. Once that quantity is consumed, the signal is sent to the supplier to replenish the quantity required.

    a. standard work rate

    b. production consumption rate

    c. postproduction sorting rate

    d. delivery requirements

21. In the 5S methodology, the Japanese word for cleaning the work area is:

    a. seiri.

    b. seiton.

    c. seiso.

    d. seiketsu.

22. A customer order for 1000 quarters is due in five days. ABC runs one eight-hour shift per day. What lean tool will help with reducing the impact of maintenance on the process?

    a. Poka-yoke

    b. SMED

    c. 5S

    d. Standard work

23. Which quality guru developed the *kanban* concept?

    a. Genichi Taguchi

    b. Masaaki Imai

    c. Taiichi Ohno

    d. W. Edwards Deming

24. Producing a punch tool that reduces costs and cycle time by creating the reeds at the same time the diameter is being punched is an example of:

    a. standard work.

    b. a kaizen event.

    c. poka-yoke.

    d. takt time.

25. A customer order for 1000 quarters is due in five days. ABC runs one eight-hour shift per day. What is the takt time?

    a. .417

    b. 2.4

    c. 25

    d. 625

26. Which of the following is another name for poka-yoke?

    a. Waste

    b. Error-proofing

    c. Orderliness

    d. Pull system

27. In a lean system, what is another name for waste?

    a. Kanban

    b. Kaizen

    c. Seiton

    d. Muda

28. The principle that each process should be done the same way, every time, by every employee is called:

    a. kanban.

    b. 5S.

    c. standardized work.

    d. pull system.

29. Which of the following tools is typically used first in a lean system?

    a. 5S

    b. Kanban

    c. Value stream map

    d. Visual control

30. In the 5S methodology, the Japanese word for sorting is:

    a. seiri.

    b. seiton.

    c. seiso.

    d. seiketsu.

31. What purpose does the material review board of a company serve?

    a. Oversee the traceability of a product

    b. Develop sampling plans for product materials

    c. Identify the source of all product components

    d. Determine corrective action for nonconforming components

32. Keeping nonconforming products from being mixed with good/conforming products within a manufacturing plant is known as:

    a. compliance.

    b. material segregation.

    c. material review.

    d. defect classification.

33. A recall due to an airbag malfunction on an automobile was issued by a car company. How would the company classify this type of defect?

    a. Minor defect

    b. Major defect

    c. Serious defect

    d. Critical defect

34. Customers have complained about the battery life of a new line of cell phones released. How would the company classify this type of product defect?

    a. Minor defect

    b. Major defect

    c. Serious defect

    d. Critical defect

Use the following information to answer questions 35–39.

A lot containing 100 printed circuit boards (PCBs) was received and sent to incoming inspection. The inspection plan called for 15 randomly selected boards to be tested.
During incoming inspection, the 15 randomly selected PCBs failed the "burn-in" test. The 15 PCBs were immediately labeled, placed on electronic hold, and moved to the quarantine storage area. A nonconformance report was initiated and placed on the agenda for the material review board.

35. What is the first action the MRB should perform?

    a. Issue a SCAR to the supplier

    b. Ensure that the remaining 85 PCBs are inspected

    c. Issue a rework order

    d. Ensure that the remaining 85 PCBs are quarantined

36. The PCBs are critical to the product being manufactured. During the MRB meeting, the incoming inspection manager reported that the last lot of this particular PCB was not inspected due to the skip-lot sampling scheme. What action should the MRB perform?

    a. Issue a SCAR to the supplier

    b. Release the remaining 85 PCBs to production

    c. Place all PCBs from the previous lot on hold

    d. Issue a rework order

37. The production manager is worried that the month's production quota will not be met and wants the MRB to release the PCBs from both lots to production and to ship the equipment that was assembled using the PCBs from the skip-lot. What is the most reasonable action for the MRB?

    a. Issue a SCAR to the supplier.

    b. Nothing, the MRB should wait for the supplier's response.

    c. Perform a retest based on an AQL sampling on the equipment that was assembled using the PCBs from the skip-lot and release.

    d. Perform a retest of all of the equipment that was assembled using the PCBs from the skip-lot and release.

38. The supplier of the PCBs informs the MRB that a new manufacturer of a component was utilized beginning with the lot of 100 PCBs that failed incoming inspection. What is the most reasonable action for the MRB?

    a. Issue a SCAR to the supplier.

    b. Nothing, the MRB should wait for the supplier's response.

    c. Perform a retest based on an AQL sampling of the equipment that was assembled using the PCBs from the skip-lot and release.

    d. Perform a retest of all of the equipment that was assembled using the PCBs from the skip-lot and release.

39. Which of the following would be an acceptable response from the supplier?

    a. Provide change notification.

    b. Have the supplier audit their supplier.

    c. Have the supplier change suppliers.

    d. Nothing, this was a one-time issue.

40. A visual identification of many potential causes of a problem is referred to as a:

    a. process flowchart.

    b. cause-and-effect diagram.

    c. decision tree.

    d. check sheet.

41. A process improvement team has completed a PFMEA. Which of the following tools would best be suited to help the team focus their efforts based on the PFMEA?

    a. Pareto analysis

    b. Force-field analysis

    c. SWOT analysis

    d. Statistical analysis

42. The causal factor or factors that, if removed, will prevent the recurrence of the same situation describe:

    a. root cause analysis.

    b. FMEA.

    c. process control.

    d. root cause.

43. A team was put together to analyze potential actions that could lead to process failures. The goal is for the team to help determine the risk associated with process failures and to prevent future failures from occurring. Which quality tool would help the team visualize this scenario?

    a. Fault tree

    b. Affinity diagram

    c. Matrix diagram

    d. Process decision program chart

A three-factor full factorial experiment was conducted. Each factor has two levels. Use the following figure to answer questions 44 and 45.

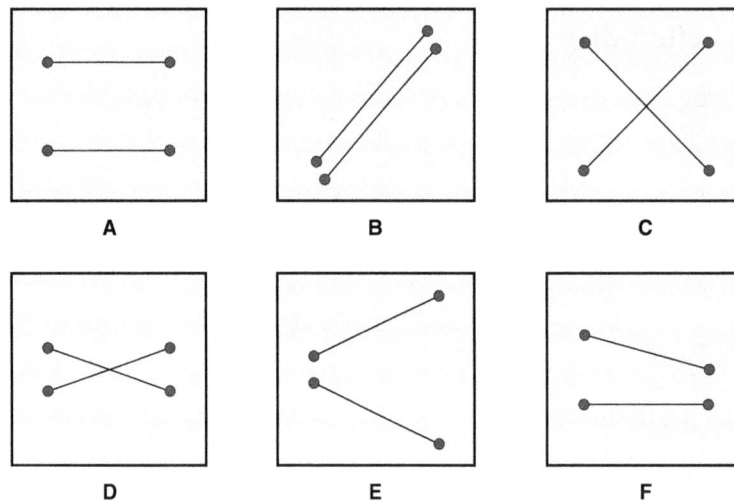

44. Which of the following plots do not indicate the presence of an interaction?

    a. A and B

    b. C and D

    c. E and F

    d. None of these

45. Which of the following plots indicate the most significant interactions?

    a. A and B

    b. B and F

    c. A and F

    d. C and D

46. A technique used by supplier quality engineers for the generation of ideas—where ideas are written down without any discussion, after which the ideas are ranked—is:

    a. brainstorming.

    b. affinity analysis.

    c. nominal group technique.

    d. configuration management.

47. The tool that can be used to organize the ideas developed during a brainstorming session into larger categories is the:

    a. interrelationship diagram.

    b. Gantt chart.

    c. affinity diagram.

    d. work breakdown structure.

48. If the bias of a measuring system increases across the measurement range, the measuring device is said to have poor:

    a. stability.

    b. linearity.

    c. bias.

    d. reproducibility.

Use the following figure to answer questions 49 and 50.

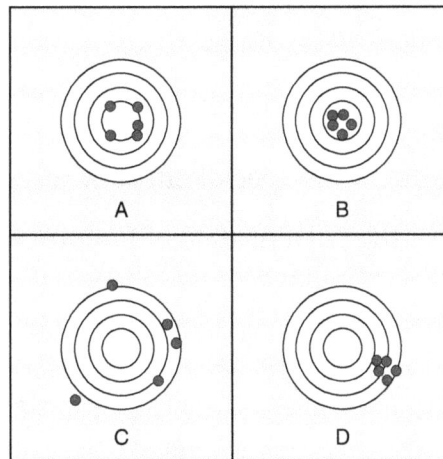

49. Which of the targets indicates good repeatability but poor bias?

    a. A

    b. B

    c. C

    d. D

50. Which of the targets indicates good repeatability, linearity, bias, and accuracy?

    a. A

    b. B

    c. C

    d. D

Use the following table to answer questions 51–54.

Row	Current process				Action results			
	Severity	Occurrence	Detection	RPN	Severity	Occurrence	Detection	RPN
1	3	5	10	150	3	5	3	45
2	5	10	3	150	3	5	3	45
3	10	3	5	150	10	3	1	30

51. On which row of the FMEA should a process improvement team focus their efforts for the current state of the process?

    a. 1

    b. 2

    c. 3

    d. The RPNs are the same, so no special effort is necessary.

52. Rank the order of risk for the current state of the process:

    a. 3, 2, 1

    b. 2, 3, 1

    c. 1, 2, 3

    d. The RPNs are the same, therefore the risk is the same.

53. What is the overall percentage of RPN reduction in row 1?

    a. 30%

    b. 45%

    c. 55%

    d. 70%

54. What is the overall percentage of RPN reduction in row 3?

    a. 20%

    b. 30%

    c. 80%

    d. 85%

55. The primary purpose of the control plan is to:

    a. keep only one item as a variable and maintain all other characteristics as controls per experiment.

    b. implement validations on specific processes.

    c. ensure that process changes are maintained over time.

    d. create structure for how quality will be implemented throughout the process.

56. Sampling for a process step would most likely be found in which column of the control plan?

    a. Characteristic

    b. Specification

    c. Frequency

    d. Reaction plan

57. What to do when the critical parameter on the control plan goes out of control is indicated on the control plan. It would most likely be found in the _____ column.

    a. reaction plan

    b. characteristic

    c. specification

    d. frequency

Use the figure below to answer questions 58 and 59.

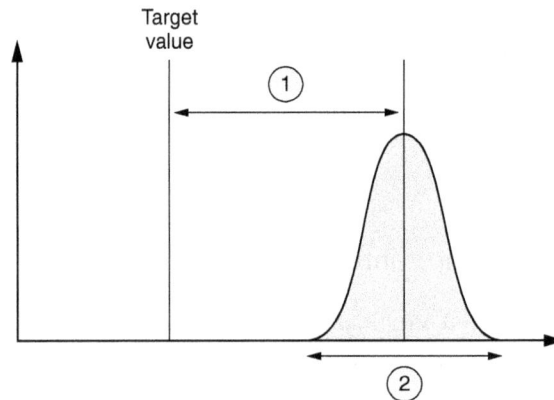

58. The encircled number 1 relates most closely to:

    a. the lower control limit.

    b. the upper control limit.

    c. accuracy.

    d. precision.

59. The encircled number 2 relates most closely to:

    a. the lower control limit.

    b. the upper control limit.

    c. accuracy.

    d. precision.

60. A tool used to monitor a process to make sure that the process improvements made are continuing to produce desired results is:

    a. the control plan.

    b. process metrics.

    c. the control chart.

    d. gage R&R.

ANSWERS

1. c;

$$\text{Acceptance } \% = \frac{26 \text{ (Lots accepted)}}{27 \text{ (Lots received)}} = 96.30\%$$

[IV.A.1]

2. a;

$$\text{Receipt } \% = \frac{1 \text{ (Lots rejected)}}{27 \text{ (Lots received)}} = 3.70\%$$

[IV.A.1]

3. c;

$$\frac{768 \text{ (Total parts)} - 17 \text{ (Rejected parts)}}{768 \text{ (Total parts)}} = 97.79\%$$

[IV.A.1]

4. b; A Pareto chart is a quality tool that can be used to organize data for analyzing trends. [IV.A.1]

5. d; The risk-based SCAR decision matrix process follows the following sequence: identify the risk severity, determine the frequency, identify the severity-frequency rating, and take appropriate action. [IV.A.1]

6. b; SCAR (supplier corrective action request). SCAR can be issued for audit findings and other supplier performance issues. [IV.A.1]

7. c; Supplier corrective action request (SCAR) is the best formal tool used for communicating quality issues with a supplier. [IV.A.1]

8. c;

$$\text{Product quality} = \frac{25 - 3}{25} \times 60 = 52.8.$$

[IV.A.1]

9. a;

$$\text{On-time delivery} = \frac{20}{25} \times 40 = 32.0.$$

[IV.A.1]

10. c; Scorecard rating = 52.8 + 32.0 = 84.8. [IV.A.1]

11. d; Lean manufacturing principles generally do not include the use of design of experiments techniques. [IV.A.2]

12. b; Single-minute exchange of die (SMED) generally has a target time of 10 minutes. [IV.A.2]

13. d; When developing standard work, the current best practice is to use photographs because visual instructions are often easier to understand and can help with training new people. [IV.A.2]

14. a; Many experts consider *safety* to be the sixth S in a 5S program. [IV.A.2]

15. c; *Kanban* is a Japanese word that means "visual signal" or "card." The signal is to perform a specific task or activity in a certain quantity. This technique is used in production and business processes that rely on a pull request or signal to do the work. [IV.A.2]

16. a; The true cost of inventory can be between 10% and 20% of the actual total dollar value at the component or piece part level, and up to 30% at the subassembly level. [IV.A.2]

17. c; The scope of work should be outlined or documented on a flowchart. [IV.A.2]

18. d; Procedures and work instructions require proper revision control as well as change management. [IV.A.2]

19. c; For general manufacturing, average inventory turns range from five to seven. Inventory on hand that is not being converted into sales is tying up cash that could be used for capital improvements or quality initiatives. [IV.A.2]

20. b; Kanban uses the production consumption rate to determine the right amount of inventory. Once that quantity is consumed, the signal is sent to the supplier to replenish the quantity required. [IV.A.2]

21. c; The 5S elements are:

- *Seiri* (Sort). Eliminate whatever is not needed

- *Seiton* (Straighten). Organize whatever remains

- *Seiso* (Shine). Clean the work area

- *Seiketsu* (Standardize). Schedule regular cleaning and maintenance

- *Shitsuke* (Sustain). Make 5S a way of life

[IV.A.3]

22. b; Single-minute exchange of die (SMED) is a system used to reduce changeover time and improve timely response to demand. SMED is a series of techniques pioneered by Shigeo Shingo to facilitate changeovers of production machinery in less than 10 minutes. [IV.A.3]

23. c; Taiichi Ohno, an industrial engineer at Toyota, developed kanban to improve manufacturing efficiency. [IV.A.3]

24. b; *Kaizen* is a Japanese term that means gradual unending improvement by doing little things better and setting and achieving increasingly higher standards. [IV.A.3]

25. b; The takt time to produce 1000 quarters due in five days with one eight-hour shift per day is calculated by

$$\text{Takt time} = \frac{\text{Time available}}{\text{Number of units to be produced}}$$

$$= \frac{8 \times 60 \times 5}{1000} = 2.4 \text{ minutes per unit}$$

[IV.A.3]

26. b; Poka-yoke comes from Japan, and is also known as *error-proofing*. It is a method of preventive action—a technique used to prevent errors from occurring. [IV.A.3]

27. d; *Muda* is another name for waste. [IV.A.3]

28. c; Standardized work means that every step or activity should be done the same way. Standardized work helps to reduce process variation and provides a more consistent product or service. [IV.A.3]

29. c; The goal of a lean system is to eliminate waste from a process. The first step is to identify opportunities for improvement, which can be found using a value stream map. Value stream maps are a visual representation of all the steps in a process, including information about timing. Value stream maps also identify the value-added and non-value-added activities in a process. [IV.A.3]

30. a; The 5S elements are:

- *Seiri* (Sort). Eliminate whatever is not needed

- *Seiton* (Straighten). Organize whatever remains

- *Seiso* (Shine). Clean the work area

- *Seiketsu* (Standardize). Schedule regular cleaning and maintenance

- *Shitsuke* (Sustain). Make 5S a way of life

[IV.A.3]

31. d; The purpose of a material review board (MRB) is to determine the appropriate corrective action to take in the case that nonconforming components are discovered. The MRB also determines any additional actions to take to prevent similar nonconformities in the future. [IV.B]

32. b; Material segregation ensures that nonconforming products are not mixed with conforming products. Segregation can either be physical, electronic using an ERP system, or a combination of both. [IV.B]

33. d; An airbag malfunction in a car could lead directly to a severe customer injury; therefore, this defect would be classified as critical. [IV.B]

34. b; The poor battery life of the cell phone reduces the usability of the product but will not likely lead to an injury or significant economic loss. Therefore, the company would classify this as a major defect. [IV.B]

35. d; The 15 randomly selected PCBs failed the "burn-in" test. These 15 PCBs were immediately labeled, placed on electronic hold, and moved to the quarantine storage area. The MRB should first ensure that the remaining 85 PCBs are placed in quarantine per procedure. [IV.B]

36. c; Because these PCBs are critical to the product being manufactured, the MRB should place all PCBs from the previous lot on hold. [IV.B]

37. d; The best course of action for the MRB would be to perform a retest of all of the equipment that was assembled using the PCBs from the skip-lot and release. [IV.B]

38. c; Because there is an assignable cause associated with the lot of PCBs that failed incoming inspection, the most reasonable action for the MRB would be to retest based on an AQL sampling of the equipment that was assembled using the PCBs from the skip-lot and release. [IV.B]

39. a; Executing a *supplier change notification* agreement is the best course of action. This would allow the company to evaluate whether the change could have a potential adverse effect. [IV.B]

40. b; A cause-and-effect diagram (also called the *Ishikawa diagram* or *fishbone diagram*) traditionally divides causes into several generic categories. In use, a large empty diagram is often drawn on a whiteboard or flip chart to visually display potential causes of a problem. [IV.C.1]

41. a; Pareto charts can be used to help a process improvement team focus their efforts based on the PFMEA. [IV.C.1]

42. d; Solving a process problem means identifying the root cause and eliminating it. The ultimate test of whether the root cause has been eliminated is the ability to toggle the problem on and off by removing and reintroducing the root cause. [IV.C.1]

43. a; Fault trees provide a visualization of hierarchical relationships of events that lead to failure (or other undesirable outcome). The team here wishes to determine potential actions that can lead to process failures. [IV.C.1]

44. a; Figures A and B do not indicate the presence of an interaction because the lines are parallel.

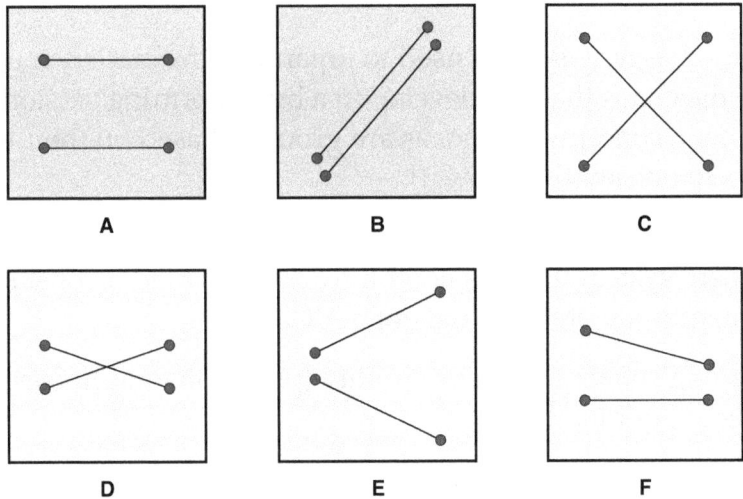

[IV.C.1]

45. d; Figures C and D indicate the most significant interactions because the lines intersect. Figures E and F also indicate the presence of interaction. However, the interaction is not as significant as in Figures C and D.

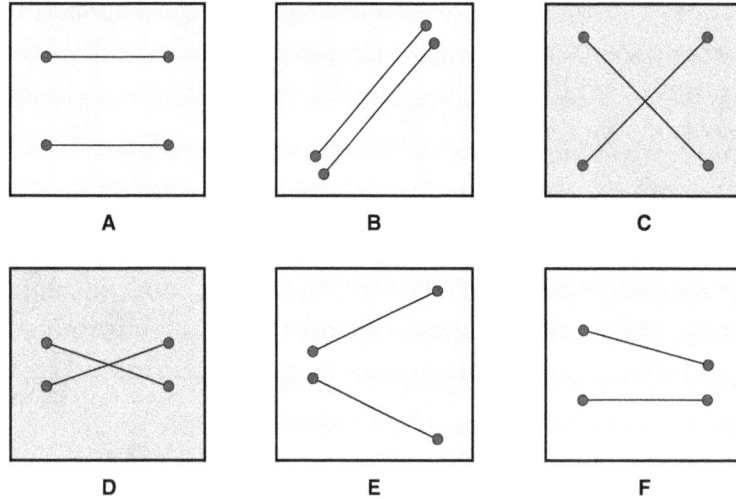

[IV.C.1]

46. a; Brainstorming is a group process used to generate ideas in a nonjudgmental way. The purpose of brainstorming is to generate a large number of ideas about an issue. [IV.C.1]

47. c; The affinity diagram is a tool used to organize information and help achieve order out of the chaos that can develop in a brainstorming session. Large amounts of data, concepts, and ideas are grouped based on their natural relationships to one another. [IV.C.1]

48. b; *Linearity* is the difference in bias through the operating range of measurements. A measurement system that has good linearity will have a constant bias no matter the magnitude of measurement. [IV.C.1]

49. d; This grouping shows good repeatability (precision) and linearity, but low accuracy due to poor bias.

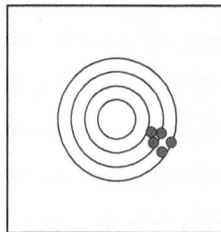

[IV.C.1]

50. b; This grouping shows good repeatability (precision), linearity, bias, and accuracy.

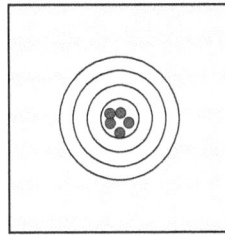

[IV.C.1]

51. c;

Row	Current process				Action results			
	Severity	Occurrence	Detection	RPN	Severity	Occurrence	Detection	RPN
1	3	5	10	150	3	5	3	45
2	5	10	3	150	3	5	3	45
3	10	3	5	150	10	3	1	30

The process improvement team should first focus their efforts on the current state of the process for row 3, as row 3 has the highest severity (10). [IV.C.2]

52. a;

Row	Current process				Action results			
	Severity	Occurrence	Detection	RPN	Severity	Occurrence	Detection	RPN
1	3	5	10	150	3	5	3	45
2	5	10	3	150	3	5	3	45
3	10	3	5	150	10	3	1	30

The order of risk for the current state of the process is row 3, row 2, and row 1. All three rows have the same RPN. Based on the information provided, the risk priority will be determined by the severity. [IV.C.2]

53. d;

Row	Current process				Action results			
	Severity	Occurrence	Detection	RPN	Severity	Occurrence	Detection	RPN
1	3	5	10	150	3	5	3	45
2	5	10	3	150	3	5	3	45
3	10	3	5	150	10	3	1	30

The percentage of RPN reduction in row 1 is calculated by the formula

$$\% \text{ RPN reduction} = \frac{\text{RPN}_i - \text{RPN}_r}{\text{RPN}_i} = \frac{150 - 45}{150} = 0.70 \text{ or } 70\%$$

where

RPN_i = Initial RPN

RPN_r = Revised RPN

[IV.C.2]

54. c;

	Current process				Action results			
Row	Severity	Occurrence	Detection	RPN	Severity	Occurrence	Detection	RPN
1	3	5	10	150	3	5	3	45
2	5	10	3	150	3	5	3	45
3	10	3	5	150	10	3	1	30

The percentage of RPN reduction in row 1 is calculated by the formula

$$\% \text{ RPN reduction} = \frac{\text{RPN}_i - \text{RPN}_r}{\text{RPN}_i} = \frac{150 - 30}{150} = 0.80 \text{ or } 80\%$$

where

RPN_i = Initial RPN

RPN_r = Revised RPN

[IV.C.2]

55. d; A control plan is a living document that identifies critical input and output variables and associated activities that must be performed to maintain control of the variation of processes, products, and services in order to minimize deviation from their preferred values. A control plan is defined as a living document; it is designed to be maintained. [IV.C.2]

56. c; The control plan should specify a sampling plan that experience has shown to be effective in detecting changes to the process. The plan should identify gages or fixtures to be used. If measurements are to be made, the plan should cross-reference a control plan for monitoring GR&R for the gage involved. [IV.C.2]

57. a; The control plan should list the steps to be taken when a process change has been detected. This serves as an aid to the responsible personnel during what is often a stressful time. The reaction plan section should cover requirements for containment and inspection of products suspected of having defects. It should also discuss disposition of parts found to be defective. Some control plans prescribe a more intense sampling protocol after certain corrective actions have been taken. [IV.C.2]

58. c; *Accuracy* refers to how closely the values are centered relative to the target/nominal value. [IV.C.2]

59. d; *Precision* refers to how closely the values are grouped or clustered together. [IV.C.2]

60. c; Once a process has been improved, it must be monitored to ensure that the gains are maintained and to determine when additional improvements are required. Control charts are used to monitor the stability of the process and determine when a special cause is present and when to take appropriate action. [IV.C.2]

Part V

Supplier Quality Management

(68 questions)

A. SUPPLIER QUALITY MONITORING

1. *Supplier audit.* Describe and distinguish between the stages of a quality audit, from audit planning through conducting the audit. Understand and apply the various types of quality audits such as product, process, and management system. (Apply)

2. *Audit reporting and follow-up.* Apply process audit reporting and follow up, including verification of the effectiveness of corrective action. (Apply)

3. *Supplier communication.* Evaluate various communication techniques such as, periodic reviews, metric and performance indices, change management, notifications, recalls, change requests and business updates. Maintain active communication with suppliers to assess risk and take appropriate action. (Evaluate)

4. *Supplier development and remediation.* Identify and analyze present and future training needs and gaps, using quality methods and tools, such as Kaizen, and benchmarking. Use process improvement tools such as, DMAIC, cycle time reduction, defect rate and cost reduction. Evaluate supplier remediation to develop and manage improvement plans. (Evaluate)

5. *Project management basics.* Understand and apply various types of project reviews, such as phase-end, management, and retrospectives or post-project reviews to assess project performance and status, to review issues and risks, and discover and capture lessons learned from the project. Apply forecasts, resources, schedules, task and cost estimates to develop and monitor project plans. (Apply)

B. TEAMS AND TEAM PROCESSES

1. *Team development.* Identify and describe the various types of teams and the classic stages of team development: forming, storming, norming, performing, and adjourning. (Apply)

2. *Team roles.* Define and describe various team roles and responsibilities for leader, facilitator, coach, and individual member. (Understand)

3. *Performance and evaluation.* Describe various techniques to evaluate training, including evaluation planning, feedback surveys, pre-training and post-training testing. (Understand)

C. COMPLIANCE WITH REQUIREMENT AND SUPPLIER CATEGORIZATION

Understand and evaluate compliance with regulations (e.g. RoHS, Governmental regulatory authorities), specifications, contracts, agreements and certification authority. Evaluate and categorize suppliers based on risk and performance. (Evaluate)

QUESTIONS

1. Which of the following would be an example of a first-party audit?

    a. An ISO registration audit

    b. A departmental audit

    c. A dock audit

    d. A supplier audit

2. Which of the following would be an example of a second-party audit?

    a. An ISO registration audit

    b. A departmental audit

    c. A dock audit

    d. A supplier audit

3. Which of the following would be an example of a third-party audit?

    a. An ISO registration audit

    b. A departmental audit

    c. A dock audit

    d. A supplier audit

Use the following information to answer questions 4–6.

> *An audit is being performed, specifically focusing on documentation (SOPs and WIs) and records. The documentation policy requires that the "same function" that originally approved the document must approve changes to documents. The procedure also requires that the documents be available to the employees electronically. The records policy requires the record to be legible, protected from unintended alterations, and with electronic scans acceptable as the permanent record. During the audit, the auditor observed the following:*
>
> > SOP 110 Preventive Maintenance—*the procedure was originally approved 10/1/2017 by Jane Doe, Maintenance Manager, and Bill Smith, Quality Manager.*
> >
> > SOP 115 Production Control—*the procedure was originally approved 11/12/2016 by Todd Jones, Production Manager, and Bill Smith, Quality Manager.*

Production Batch Record 1234 was scanned and saved on the company server, which is cloud based. The server requires a unique user name and password for access. The batch was approved for release on 3/14/2018.

4. *SOP 110 Preventive Maintenance* needs to be updated; however, Jane Doe, Maintenance Manager, was promoted to director of operations, and a replacement was recently hired. Which of the following individual(s) should approve the change to the procedure?

    a. Only the quality manager because he was an original signer

    b. The quality manager and the maintenance manager

    c. The VP of operations and the maintenance manager

    d. The VP of operations, the maintenance manager, and the quality manager

5. *SOP 115 Production Control Documentation* needs to be updated; however, Todd Jones, Production Manager, is on vacation for six weeks. The production manager directly reports to the VP of operations. Which of the following individual(s) should approve the change to the procedure?

    a. The VP of operations and the quality manager.

    b. The VP of operations and the maintenance manager.

    c. Only the quality manager because he was an original signer.

    d. The procedure should not be changed until production manager Todd Jones returns from vacation.

6. The auditor noticed that Production Batch Record 1234 was approved for release on 3/14/2018, scanned, and saved to the cloud-based server. The auditor asked for the original paper copy. The VP of operations informed the auditor that under the procedure, scanned electronic copies are acceptable as the permanent record. Is this a violation of the procedure?

    a. Yes, original paper documents should never be destroyed.

    b. Yes, the cloud-based server is not physically located on the company property.

    c. No, paper copies are required to be saved.

    d. No, the procedure allows electronic scans as the permanent record.

7. What type of audit evaluates final "fitness for use"?

    a. Product audit

    b. System audit

    c. Process audit

    d. Desk audit

8. What type of audit is best suited for continuous improvement?

    a. Product audit

    b. System audit

    c. Process audit

    d. Desk audit

9. What type of audit is best suited to verify the adequacy of the QA program?

    a. Product audit

    b. System audit

    c. Process audit

    d. Desk audit

10. Which of the following is the weakest reason for a quality audit?

    a. Disciplinary control

    b. Product quality evaluation

    c. Preventive action

    d. Qualifying a new supplier

11. Which of the following ISO standards provides guidance on auditing management systems?

    a. ISO 9001

    b. ISO 13485

    c. ISO 15378

    d. ISO 19011

12. The _____ is responsible for the audit report's content, accuracy, and timely publication.

    a. audit team

    b. sourcing/purchasing lead

    c. lead auditor

    d. supplier quality manager

Part V
Questions

13. If the gap between an audit and the audit report is excessive, the auditee might get the unintended message that:

    a. auditors are overwhelmed.

    b. the audit and the corrective action on its findings are not a priority.

    c. the audit was a "paper" exercise.

    d. they did really well during the audit.

14. Audit reports should not include:

    a. argumentative statements.

    b. subjective opinions.

    c. proprietary information.

    d. all of the above.

15. Audit records include the following *except*:

    a. auditor training and credentials.

    b. auditors' notes.

    c. the audit plan.

    d. records of corrective and preventive actions.

16. A supplier corrective action can be considered closed when:

    a. the plan is submitted to the customer.

    b. purchasing and supplier quality approve.

    c. it has been implemented and verified.

    d. filed with the auditor.

17. The best person to approve a supplier audit response is:

    a. the supplier quality manager.

    b. the director of procurement.

    c. the supplier's management team.

    d. the auditor.

18. Strategies for optimizing overall quality cost is a topic for:

    a. joint economic planning.

    b. quarterly business review.

   c. a supplier scorecard.

   d. management review.

19. Total dollar value purchased (spend), percent defective dollars versus spend, percent of lots rejected, and composite supplier rating scores are all examples of:

   a. a quarterly business review (QBR) agenda.

   b. financial due diligence.

   c. supplier feedback topics.

   d. supplier selection criteria.

Questions 20 through 24 concern different types of supplier assessment metrics. Choose the category based on the list offered.

20. Percent on-time delivery and percent late deliveries are examples of:

   a. timeline metrics.

   b. quality metrics.

   c. delivery (JIT) metrics.

   d. compliance metrics.

21. Percent over-orders, percent under-ordered, and early delivery are examples of:

   a. timeline metrics.

   b. quality metrics.

   c. delivery (JIT) metrics.

   d. compliance metrics.

22. Percent supplies missing certificates, percent of reported required quality information are examples of:

   a. timeline metrics.

   b. quality metrics.

   c. delivery (JIT) metrics.

   d. compliance metrics.

23. Percent defective, ppm, DPMO, percent nonconforming are examples of:

   a. cost metrics.

   b. quality metrics.

   c. delivery (JIT) metrics.

   d. compliance metrics.

24. Dollars rejected versus purchased is an example of a:

   a. cost metric.

   b. quality metric.

   c. delivery (JIT) metric.

   d. compliance metric.

25. Incremental continuous improvement is also known as:

   a. kaizen.

   b. Six Sigma.

   c. lean enterprise.

   d. plan–do–check–act.

26. The "C" in DMAIC is critical. Why?

   a. It stands for *cost* and serves as justification for the entire project.

   b. It stands for *control*, specifically of costs and employee behavior.

   c. It stands for *control*, and is vital to maintain the gains achieved by the project.

   d. It's not any more critical than any other phase of a project.

27. Excess motion, repairs/rework, waiting, and excess inventory are some examples of:

   a. steps in a process.

   b. muda.

   c. cycle time reduction.

   d. steps to take to ensure material is ready when production is.

28. Efforts to improve processes to reduce waste, increase throughput, and increase capacity are all indicative of:

   a. cycle time reduction.

   b. reductions in workforce.

   c. the company being sold.

   d. good leadership.

29. Supplier development is an endeavor that adds value:

    a. by early involvement of the supplier quality group during the selection process.

    b. by partnering with a supplier to raise the level of quality and service performance.

    c. by developing strong personal relationships with the supplier's key contacts.

    d. by developing new core competencies with existing suppliers (for example, teaching molders to extrude).

30. Supplier remediation activities include:

    a. PPAP instruction.

    b. joint kaizen events.

    c. technical assessments.

    d. all of the above.

31. Supplier remediation usually occurs:

    a. at the qualification phase.

    b. after performance issues arise.

    c. as a result of the supplier's internal audit.

    d. after a regulatory inspection.

32. SMART goals—This method helps ensure that the goals have been fully investigated and provides a way to clearly understand the implications of the goal-setting process. What does the M refer to?

    a. Meaningful

    b. Management

    c. Metrics

    d. Measurable

33. CLEAR goals—A newer method for setting goals that takes into consideration the environment of today's fast-paced business. What does the L refer to?

    a. Lesser

    b. Larger

    c. Limited

    d. Leveraged

34. What type of chart provides a visual timeline that can be used to plan tasks and visualize the project timeline.

    a. Pareto chart

    b. Gantt chart

    c. Flowchart

    d. Control chart

35. What type of plan is used to identify unrealistic time and cost estimates, customer review cycle, budget cuts, changing requirements, and lack of committed resources?

    a. Risk management

    b. Communication

    c. Budget

    d. Project

36. _____ is a visual representation that breaks down the project scope into manageable sections for the team.

    a. DMAIC

    b. PDCA

    c. WBS

    d. DOE

37. _____ identify high-level goals that need to be met throughout the project. These are included in the Gantt chart.

    a. Metrics

    b. Measurables

    c. Specifications

    d. Milestones

38. Goals are normally set in the planning stage. Which of the following tools is best used to determine project goals?

    a. PDCA

b. CLEAR

c. SMART

d. Both (b) and (c)

39. The team phase where team members struggle to understand the team goal and its meaning for them individually is referred to as:

a. forming.

b. storming.

c. norming.

d. performing.

40. The team phase where team members begin to understand the need to operate as a team rather than as a group of individuals is referred to as:

a. forming.

b. storming.

c. norming.

d. performing.

41. The team phase where team members express their own opinions and ideas, often disagreeing with others, is referred to as:

a. forming.

b. storming.

c. norming.

d. performing.

42. The team phase where team members understand one another and recognize each other's strengths and weaknesses is referred to as:

a. performing.

b. forming.

c. storming.

d. norming.

43. A high-performing team has recently had changes to the individuals participating in the team project. With the change in personnel, in what stage is the team likely to find itself?

    a. Storming

    b. Forming

    c. Norming

    d. Performing

44. What are the five stages of team development, in order?

    a. Performing, storming, norming, adjourning, forming

    b. Storming, adjourning, norming, forming, performing

    c. Forming, storming, norming, performing, adjourning

    d. Storming, performing, forming, norming, adjourning

45. How can a team best overcome anxiety during the forming stage?

    a. Create a team name and slogan.

    b. Let members pick their own teams.

    c. Determine a team leader quickly.

    d. Set clear goals, timelines, and roles to communicate expectations.

46. This audit team role is responsible for calling the meetings, making meeting arrangements, running the meetings, and reporting progress to the client and the auditee.

    a. Client

    b. Team members

    c. Facilitator

    d. Lead auditor

47. _____ may include qualified auditors, auditors in training, and technical experts.

    a. Client

    b. Audit team members

    c. Facilitator

    d. Lead auditor

48. This role assists the lead auditor in organizing the team and making it more effective, but does not participate as a team member.

   a. Client

   b. Auditor

   c. Facilitator

   d. Lead auditor

49. This audit team role ensures that everyone understands his or her assignment before the audit starts.

   a. Client

   b. Auditor

   c. Facilitator

   d. Lead auditor

50. This audit team role should provide the necessary resources for the team members, such as forms, examples, and working papers, and if necessary instruct team members on the use of working papers.

   a. Client

   b. Auditor

   c. Facilitator

   d. Lead auditor

51. Audit teams are _____ teams formed to carry out the audit plan and achieve the supplier audit objectives.

   a. ad hoc

   b. administrative

   c. specialized

   d. complicated

52. The need for facilitators and facilitation techniques _____ with team member training, experience, and successful team history.

   a. increases

   b. decreases

   c. remains the same

   d. is not considered

53. A situation in which critical information is withheld from the team because individual members censor or restrain themselves, either because they believe their concerns are not worth discussing or because they are afraid of confrontation.

    a. Storming

    b. Socializing

    c. Brainstorming

    d. Groupthink

54. The team facilitator asks each team member to suggest one idea during a brainstorming session. This method can be used when:

    a. the meeting session is over two hours.

    b. the team is behind schedule.

    c. some team members are dominating the discussion.

    d. management is observing the team.

55. Which of the following is a characteristic of a team that is effectively functioning?

    a. A sense of team interdependence

    b. A clear sense of purpose

    c. A focus on individual tasks

    d. A knowledge of company policies

56. A team facilitator can promote positive team relationships by:

    a. recording meeting notes.

    b. debating each item.

    c. keeping track of time.

    d. encouraging participation.

57. What can occur when team members put more value on getting along than on achieving team goals?

    a. Groupthink

    b. Conflict

    c. Synergy

    d. Collaboration

58. Which characteristic best indicates a team member's ability to contribute to a project team?

    a. Leadership skills

    b. Ability to conform

    c. Knowledge of the problem

    d. Social justification

59. A facilitator can turn team conflict into a problem-solving process by:

    a. discussing conflict resolution techniques.

    b. developing rules for conflict resolution.

    c. identifying the root cause analysis.

    d. evaluating concerns.

60. Which of the following European Union (EU) regulations restricts the use of certain chemicals, including lead (Pb), mercury (Hg), and cadmium (Cd), in electrical and electronic products?

    a. REACH

    b. RoHS

    c. OSHA

    d. Conflict Minerals

61. Registration has been adopted to meet the EU requirements to protect human health and the environment and to prevent risk posed by chemicals. This registration is known as:

    a. REACH

    b. RoHS

    c. OSHA

    d. Conflict Minerals

62. This is the most common type of specification. It defines what is required for a product to perform as expected by the consumer.

    a. Quality management specifications

    b. Product specifications

    c. Process specifications

    d. Raw material specifications

63. This type of specification defines the parameters of the manufacturing process that must be controlled to produce a product.

    a. Quality management specifications

    b. Product specifications

    c. Process specifications

    d. Raw material specifications

64. Where applicable, this type of specification defines the analytical methodologies for measuring a required level of accuracy.

    a. Quality management specifications

    b. Product specifications

    c. Analytical specifications

    d. Raw material specifications

65. This type of specification defines what is acceptable as raw material entering a manufacturing process.

    a. Quality management specifications.

    b. Product specifications.

    c. Analytical specifications.

    d. Raw material specifications.

66. This type of specification defines the management practices under which you wish to have products produced.

    a. Product specifications

    b. Quality management specifications

    c. Analytical specifications

    d. Raw material specifications

67. _____ controls the export and import of defense-related articles and services on the United States Munitions List (USML).

    a. DMAIC

    b. REACH

    c. ITAR

    d. Conflict Minerals

68. Conflict minerals are natural minerals extracted in a conflict zone and sold to perpetuate the fighting. Conflict minerals are associated with which geographic region?

    a. Middle East

    b. Antarctica

    c. North America

    d. Democratic Republic of the Congo

ANSWERS

1. b; A first-party audit is performed within an organization to measure its strengths and weaknesses against its own procedures or methods and/or against external standards adopted by (voluntary) or imposed on (mandatory) the organization. A first-party audit is an internal audit conducted by auditors who are employed by the organization being audited but who have no vested interest in the audit results of the area being audited. [V.A.1]

2. d; A second-party audit is an external audit performed on a supplier by a customer or by a contracted organization on behalf of a customer. A contract is in place, and the goods or services are being, or will be, delivered. Second-party audits are subject to the rules of contract law, as they are providing contractual direction from the customer to the supplier. Second-party audits tend to be more formal than first-party audits because audit results could influence the customer's purchasing decisions. [V.A.1]

3. a; A third-party audit is performed by an audit organization independent of the customer–supplier relationship and free of any conflict of interest. Independence of the audit organization is a key component of a third-party audit. Third-party audits may result in certification, registration, recognition, an award, license approval, a citation, a fine, or a penalty issued by the third-party organization or an interested party. [V.A.1]

4. b; *SOP 110 Preventive Maintenance* needs to be updated; however, Jane Doe, Maintenance Manager, was promoted to director of operations, and a replacement was recently hired. The procedure was originally approved 10/1/2017 by maintenance manager Jane Doe and Bill Smith, Quality Manager. The procedure should be signed by the same function, not necessarily the same individual. Therefore, the maintenance manager and the quality manager should both approve the changes to the procedure. [V.A.1]

5. a; *SOP 115 Production Control Documentation* needs to be updated; however, Todd Jones, Production Manager, is on vacation for six weeks. Because the production manager directly reports to the VP of operations, it would be appropriate for the VP of operations and the quality manager to approve the changes to the document. [V.A.1]

6. d; The auditor noticed that Production Batch Record 1234 was approved for release on 3/14/2018, scanned, and saved to the cloud-based server. The auditor asked for the original paper copy. The VP of operations informed the auditor that under the procedure, scanned electronic copies are acceptable as the permanent record. This would not constitute a violation of the procedure. However, if the scan copy were not legible, it could be a violation of the procedure. [V.A.1]

7. a; A *product audit* is an examination of a particular product or service (hardware, processed material, software) to evaluate whether it conforms to requirements (that is, specifications, performance standards, and customer requirements). [V.A.1]

8. c; A *process audit* is a verification that processes are working within established limits. It evaluates an operation or method against predetermined instructions or standards to measure conformance to these standards and the effectiveness of the instructions. [V.A.1]

9. b; A system audit is an audit conducted on a management system. It can be described as a documented activity performed to verify, by examination and evaluation of objective evidence, that applicable elements of the system are appropriate and effective and have been developed, documented, and implemented in accordance and in conjunction with specified requirements. [V.A.1]

10. a; Disciplinary control should never be the reason for a quality audit. Quality audits are used to assess systems, processes, and products, and seek improvement opportunities. [V.A.1]

11. d; ISO 19011:2011 provides guidance on auditing management systems, including the principles of auditing, managing an audit program, and conducting management system audits, as well as guidance on the evaluation of the competence of individuals involved in the audit process, including the person managing the audit program, auditors, and audit teams. [V.A.2]

12. c; The lead auditor is ultimately responsible for the audit reporting, including accuracy, completeness, and timely publication. [V.A.2]

13. b; Any findings important enough to document during an audit are also urgent to resolve. Untimely reporting sends the message that there are "more important and urgent" priorities. [V.A.2]

14. d; None of these are acceptable in an audit report. Auditors must strive for objectivity and to be factual. Since the report might be read by others, confidentiality (proprietary information) must be safeguarded. [V.A.2]

15. b; Auditor notes are working materials and are private to the auditor, as the positions expressed in the notes may evolve during the audit. They are therefore not usually considered audit records. (A distinction must be made with regard to a regulatory inspection; the notes that FDA, for example, documents *are* part of the investigation. This is one important difference between audits and inspections.) [V.A.2]

16. c; Closure can only occur after implementation *and* verification. Approval of plans or just customer approval are not enough. Objective evidence of implementation is required. A follow-up audit might be a good idea as well to verify effectiveness. [V.A.2]

17. d; Only the auditor has the in-depth first-person knowledge to appropriately judge the acceptability of an audit finding response from the supplier. [V.A.2]

18. a; Quality costs strategy is a staple of joint economic planning, along with discussing the value proposition. Actual cost of poor quality is a great category for scorecards and a topic for a QBR. [V.A.3]

19. c; These are all excellent topics to include on supplier feedback reports to enhance a strong customer–supplier relationship. They could be QBR agenda items as well, but in the end, QBR is just one type of supplier feedback. [V.A.3]

20. a; Although timeliness and delivery are similar, these are timeliness metrics. They deal with receipts of shipments. [V.A.3]

21. c; Although timeliness and delivery are similar, these are delivery metrics. They deal with just-in-time delivery. [V.A.3]

22. d; Although quality and compliance are similar, these are compliance metrics because they refer to required documentation. [V.A.3]

23. b; Although quality and compliance are similar, these are quality metrics because they refer to product quality and meeting/not meeting requirements. [V.A.3]

24. a; Because dollar value is mentioned, cost is the right choice. [V.A.3]

25. a; Kaizen refers to incremental improvement that never stops. The other choices are great improvement tools, but are not necessarily used for iterative, ongoing improvement. [V.A.4]

26. c; The C stands for *control* and is considered, along with a great problem statement, the most critical phase because many organizations make improvements but lose them over time to culture, entropy, employee behavior, or "tribal knowledge." It has nothing to do with cost control. [V.A.4]

27. b; These are verbatim examples of waste, or *muda*. The other answers are simply wrong and would show a need to conceptually revisit not only muda but the other answers as well, if selected. [V.A.4]

28. a; Cycle time is the only logical answer. It's not good leadership, and selling the company has no bearing. [V.A.4]

29. b; (a), (b), and (d) are all good answers, but (b) provides the most specific answer. [V.A.4]

Part V
Answers

30. d; All of these activities are great areas in which to resolve performance issues with the supplier. [V.A.4]

31. b; Unless they are a special supplier like an API manufacturer, most suppliers will not be subject to regulatory inspections. Although remediation can be done, and should be done up front with poor audit results, (b) is a more comprehensive answer covering poor performance at any time or phase in the relationship. [V.A.4]

32. d;

S	Specific
M	Measurable
A	Attainable
R	Realistic
T	Timely

[V.A.5]

33. c;

C	Collaborative
L	Limited
E	Emotional
A	Appreciable
R	Refinable

[V.A.5]

34. b; Gantt charts provide a graphical timeline that can be used to plan tasks and visualize the project timeline. [V.A.5]

35. a; A risk management plan can be used to identify all potential risks. These include unrealistic time and cost estimates, customer review cycle, budget cuts, changing requirements, and lack of committed resources. [V.A.5]

36. c; Work breakdown structure (WBS) is a visual representation that breaks down the project scope into manageable sections for the team. [V.A.5]

37. d; The milestones identify high-level goals that need to be met throughout the project. These are included in the Gantt chart. [V.A.5]

38. c; SMART and CLEAR tools can be used to determine project goals. [V.A.5]

39. a; Team forming is when members struggle to understand the team goal and its meaning for them individually. [V.B.1]

40. c; Team norming is when members begin to understand the need to operate as a team rather than as a group of individuals. [V.B.1]

41. a; Team storming is when members express their own opinions and ideas, often disagreeing with others. [V.B.1]

42. a; Team performing is when team members work together to reach their common goal. [V.B.1]

43. b; Team forming is when members struggle to understand the team goal and its meaning for them individually. [V.B.1]

44. c; The five stages of team development (in order) are forming, storming, norming, performing, and adjourning. [V.B.1]

45. d; Setting clear goals, timelines, and roles can help a team overcome anxiety during the forming stage. [V.B.1]

46. d; The lead auditor is responsible for calling the meetings, making meeting arrangements, running the meetings, and reporting progress to the client and the auditee. The lead auditor should take ownership of the audit process. The lead auditor normally controls the team output or deliverables. Lead auditors must be able to guide the team and make decisions as needed to ensure that the team is effective. The lead auditor should be a qualified (competent) auditor. [V.B.2]

47. b; Team members may include qualified auditors, auditors in training, and technical experts. Team members accept assignments and report progress to the lead auditor or the team as a whole. Team members may be given responsibilities such as keeping records or taking meeting minutes. [V.B.2]

48. c; A team may have a facilitator who assists the lead auditor in organizing the team and making it more effective but does not participate as a team member. In the cross-functional team for a specific purpose and the long-term project team, the facilitator or coach is included as needed. The facilitator's role is to help the team resolve issues and reach its purpose or achieve its goals effectively. [V.B.2]

49. d; The lead auditor ensures that everyone understands his or her assignment before the audit starts. [V.B.2]

50. d; The lead auditor should provide the necessary resources for the team members, such as forms, examples, and working papers, and if necessary, instruct team members on the use of the working papers. [V.B.2]

51. a; Audit teams are ad hoc teams formed to carry out the audit plan and achieve the supplier audit objectives. [V.B.2]

52. b; The need for facilitators and facilitation techniques decreases with team member training, experience, and successful team history. [V.B.2]

53. d; Groupthink is a situation in which critical information is withheld from the team because individual members censor or restrain themselves, either because they believe their concerns are not worth discussing or because they are afraid of confrontation. [V.B.3]

54. c; When a team member(s) is/are dominating a discussion, the team facilitator can ask each team member to suggest one idea during a brainstorming session. This will help solicit participation from the larger group. [V.B.3]

55. b; A characteristic of a team that is effectively functioning is a clear sense of purpose. [V.B.3]

56. d; By encouraging participation, the team facilitator can promote positive team relationships. [V.B.3]

57. a; Groupthink can occur when team members put more value on getting along than on achieving team goals. [V.B.3]

58. c; Knowledge of the problem is an indicator of a team member's ability to contribute to a project team. [V.B.3]

59. a; A facilitator can turn team conflict into a problem-solving process by discussing conflict resolution techniques. [V.B.3]

60. b; The Restriction of Hazardous Substances (RoHS) Directive restricts the types of hazardous materials that can be used in electrical and electronic products. Under RoHS, the following substances are prohibited from entering the market for electrical and electronic products: lead (Pb), mercury (Hg), cadmium (Cd), hexavalent chromium (Cr VI), polybrominated biphenyls (PBB), polybrominated diphenyl ethers (PBDE), and four different phthalates (DEHP, BBP, DBP, and DIBP). [V.C]

61. a; Registration, Evaluation, Authorization and Restriction of Chemicals (REACH) has been adopted to protect human health and the environment and to prevent risk posed by chemicals. [V.C]

62. b; Product specifications are the most common type of specification. They define what is required for a product to perform as expected by the consumer. [V.C]

63. c; Process specifications define the parameters of the manufacturing process that must be controlled in order to produce a product. [V.C]

64. c; Analytical specifications define the analytical methodologies for measuring a required level of accuracy. [V.C]

65. d; Raw material specifications define what is acceptable as raw material entering a manufacturing process. [V.C]

Part V
Answers

66. b; Quality management specifications define the management practices under which you wish to have products produced. [V.C]

67. c; International Traffic in Arms Regulations (ITAR) controls the export and import of defense-related articles and services on the United States Munitions List (USML). ITAR does not apply to general scientific, mathematical, or engineering principles that are taught in schools or are in the public domain. [V.C]

68. d; Conflict minerals are natural minerals extracted in a conflict zone and sold to perpetuate the fighting. Conflict minerals are associated with the Democratic Republic of the Congo. The four most commonly mined conflict minerals are cassiterite (for tin), wolframite (for tungsten), coltan (for tantalum), and gold ore. [V.C]

Part VI

Relationship Management

(37 questions)

A. SUPPLIER ONBOARDING

Understand and apply processes for orientation of suppliers such as, providing overview of company, vision, mission, guiding principles, overall requirements, expectations, and criticality of product, service, and delivery requirements. (Apply)

B. COMMUNICATION

1. *Techniques and mediation.* Identify and apply communication techniques (oral, written, and presentation) specifically for internal stakeholders and suppliers to resolve issues. Apply different techniques when working in multi-cultural environments, and identify and describe the impact that culture and communications can have on quality. (Evaluate)

2. *Reporting using quality tools.* Use appropriate technical and managerial reporting techniques, including the seven classic quality tools (Pareto charts, cause and effect diagrams, flowcharts, control charts, check sheets, scatter diagrams, and histograms) for effective presentation and reporting. (Analyze)

C. LEADERSHIP AND COLLABORATION

Understand and apply techniques for coaching suppliers through regular communications, influencing without authority, negotiation techniques and establish clear roles and responsibilities of internal stakeholders and suppliers. (Evaluate)

QUESTIONS

1. Which phase of the supplier orientation process is used to gather work requirements and methods, prepare work requirements for suppliers per scope of work, and establish expectations for suppliers?

   a. Prototyping

   b. Planning

   c. Development

   d. Execution

2. Which phase of the supplier orientation process is used to communicate the requirements, means and methods, schedules, and deliverables per scope of work?

   a. Prototyping

   b. Planning

   c. Development

   d. Execution

3. Which phase of the supplier orientation process is used to identify resources, processes, and work instructions to be shared with suppliers?

   a. Prototyping

   b. Planning

   c. Development

   d. Execution

4. One objective of supplier orientation is to:

   a. impart an understanding of the key work processes and requirements.

   b. introduce all the QA personnel and suppliers to each other.

   c. show the supplier your systems.

   d. give the supplier a chance to showcase their business to management.

5. One key outcome of supplier orientation is that:

   a. it creates happy suppliers.

   b. it builds the customer–supplier relationship.

    c. it lends visibility to expectations and the criticality of the outsourced work in the the customer's eyes.

    d. the supplier and customer both know more about each other's businesses and processes.

6. _____ is/are another aspect of supplier orientation.

    a. Customer satisfaction

    b. Piece part price agreement

    c. Documentation of customer–supplier disagreements

    d. Clearly defined roles and responsibilities

7. _____ is a very important part of the orientation process.

    a. Cooperation

    b. Feedback

    c. Collaboration

    d. Good communication

8. This type of communicator is a good listener, avoids displays of power, and uses collaborative methods to find solutions to problems.

    a. Emotive

    b. Director

    c. Reflective

    d. Supportive

9. An individual who is an easygoing, down-to-earth quality manager who leads by example, has a friendly demeanor, and gets people to follow him or her without difficulty is what type of communicator?

    a. Emotive

    b. Director

    c. Reflective

    d. Supportive

10. This type of communicator expresses themself in a more deliberate manner, may seem more quiet or introverted, and make decisions more slowly.

    a. Emotive

  b. Director

  c. Reflective

  d. Supportive

11. A person with a stern, serious attitude, strong opinions, and air of indifference are typical characteristics of this type of communicator.

  a. Emotive

  b. Director

  c. Reflective

  d. Supportive

12. While conducting global business transactions, cultural differences between the sender and receiver can have a significant bearing on the effectiveness of communication. Which of the following types of communication is most effective in this situation?

  a. E-mail

  b. Blog post

  c. Letter

  d. Presentation

13. A person from North America would be mostly likely have which of the following communication styles?

  a. Low-context and sequential

  b. Low-context and synchronic

  c. High-context and sequential

  d. High-context and synchronic

14. A person from Asia would be mostly likely have which of the following communication styles?

  a. Low-context and sequential

  b. Low-context and synchronic

  c. High-context and sequential

  d. High-context and synchronic

15. A process improvement team has completed a PFMEA. Which of the following tools would be best suited to help the team focus their efforts based on the PFMEA?

    a. SWOT analysis

    b. Pareto analysis

    c. Force-field analysis

    d. Statistical analysis

16. The phrase "vital few and useful many" is applicable to the:

    a. cause-and-effect diagram.

    b. check sheet.

    c. Pareto diagram.

    d. process flow diagram.

Use the following figures to answer questions 17–20.

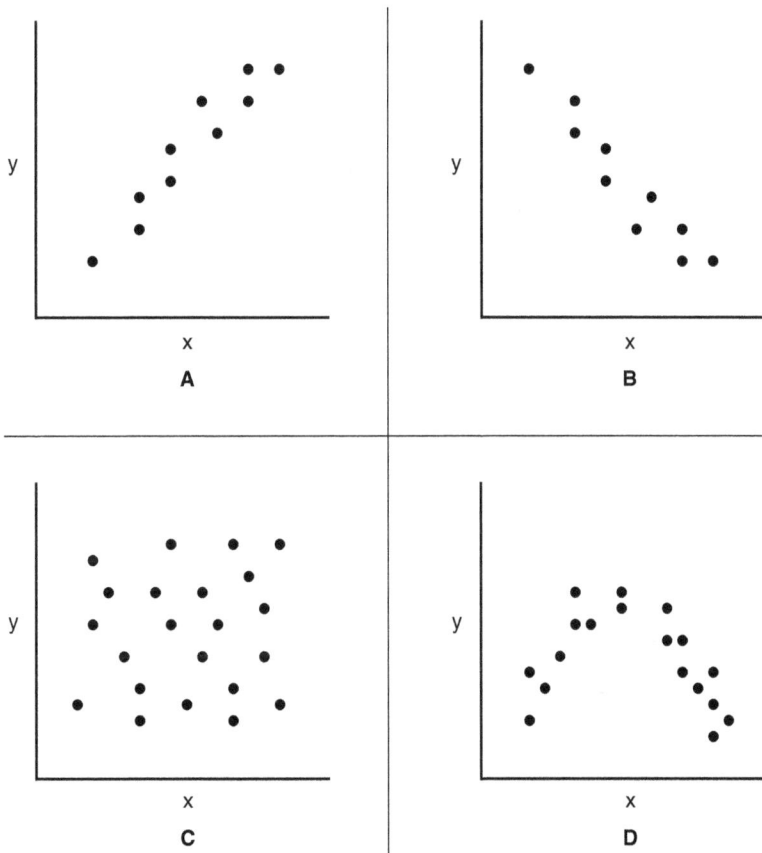

17. Which figure demonstrates negative correlation?

    a. Figure A

    b. Figure B

    c. Figure C

    d. Figure D

18. Which figure demonstrates nonlinear correlation?

    a. Figure A

    b. Figure B

    c. Figure C

    d. Figure D

19. Which figure demonstrates no correlation?

    a. Figure A

    b. Figure B

    c. Figure C

    d. Figure D

20. Which figure demonstrates positive correlation?

    a. Figure A

    b. Figure B

    c. Figure C

    d. Figure D

21. A tool used to provide visual identification of many potential causes of a problem is the:

    a. decision tree.

    b. process flowchart.

    c. cause-and-effect diagram.

    d. check sheet.

22. The tool that captures requirements on inputs into and outputs from a process is:

    a. requirements tree analysis.

    b. SIPOC.

   c. quality function deployment.

   d. the Kano model.

23. For the normal curve, approximately 68% of the data will fall between:

   a. $\pm 2\sigma$.

   b. $\pm 3\sigma$.

   c. $\pm 1\sigma$.

   d. $\pm 5\sigma$.

24. The tool used to monitor a process to make sure that the process improvements made are continuing to produce desired results is:

   a. a control chart.

   b. gage R&R.

   c. process metrics.

   d. a control plan.

25. Control charts are graphical tools used to monitor a process. Control charts generally do not contain or use:

   a. a centerline.

   b. control limits.

   c. specification limits.

   d. subgroups.

26. The SPC tool that does not have a fixed form is the:

   a. control chart.

   b. scatter diagram.

   c. check sheet.

   d. cause-and-effect diagram.

27. Which quality tool is used to diagram a process?

   a. Control chart

   b. Measles chart

   c. Venn diagram

   d. Flowchart

Part VI Questions

28. A visual identification of many potential causes of a problem is referred to as a:

    a. process flowchart.

    b. cause-and-effect diagram.

    c. decision tree.

    d. check sheet.

29. Which of the following continuous improvement methods focuses on a business's customers, quality built into the work culture, and strong leadership involvement in quality efforts?

    a. Kaizen

    b. TQM

    c. PDCA

    d. Theory of constraints (TOC)

30. Armand V. Feigenbaum's concept of total quality includes:

    a. quality leadership, modern quality technology, and organizational commitment.

    b. commitment to quality, management commitment, and measurement of potential quality problems.

    c. statistical process control.

    d. benchmarking and reengineering.

31. In which role in the RACI model must the decision be discussed with the individual before a decision is made?

    a. Responsible

    b. Accountable

    c. Consulted

    d. Informed

32. In which role in the RACI model must the individual be cognizant about a decision because they are affected?

    a. Responsible

    b. Accountable

    c. Consulted

    d. Informed

33. In which role in the RACI model is the individual ultimately held responsible for results?

    a. Responsible

    b. Accountable

    c. Consulted

    d. Informed

34. In which role in the RACI model do the individuals actively participate in an activity?

    a. Responsible

    b. Accountable

    c. Consulted

    d. Informed

35. The type of leader who provides clear direction to team members is:

    a. supportive.

    b. coaching.

    c. situational.

    d. directional.

36. Empowering process improvement team members to function independently is known as _____ leadership.

    a. situational

    b. supportive

    c. deductive

    d. directional

37. What type of leadership is best suited for high-performing team members that are technically competent?

    a. Delegating

    b. Directional

    c. Supportive

    d. Adversarial

ANSWERS

1. c; The phase of the supplier orientation process used to gather work requirements and methods, prepare work requirements for suppliers per scope of work, and establish expectations for suppliers is the *development* phase. [VI.A]

2. d; The phase of the supplier orientation process used to communicate the requirements, means and methods, schedules, and deliverables per scope of work is the *execution* phase. [VI.A]

3. b; The phase of the supplier orientation process used to identify key resources, identify processes, work instructions, and methods to be shared with suppliers, set up communication with supplier contacts, and schedule orientation sessions is the *planning* phase. [VI.A]

4. a; Although (b), (c), and (d) are all important to accomplish during orientation, the real objective is for the supplier to understand processes and requirements. [VI.A]

5. c; (a) is desirable but not the answer to this question. (b) and (d) are also good answers, but as with (4) above, it's really important for the work and requirements to be understood from the customer point of view. [VI.A]

6. d; (a) and (b) are incorrect. (c) is a part of "feedback" but not the best answer. Roles and responsibilities are key to a good relationship and smooth operation going forward. [VI.A]

7. b; (a), (c), and (d) are all important attitudes and behaviors during the orientation process, but *feedback* is an actual step in the process. [VI.A]

8. d; A supportive communicator possesses a low degree of dominance and a high degree of sociability. This person is a good listener, avoids displays of power, and uses collaborative methods to find solutions to problems. An example of a supportive communicator is a Six Sigma Black Belt practitioner who leads open, low-stress meetings, encourages participation, and facilitates finding solutions that can be supported by all of those involved. [VI.B.1]

9. a; An *emotive* communicator shows high degrees of dominance and sociability. This person typically displays action-oriented behavior, prefers informality, and is adept at persuasion. An example of an emotive communicator is an easygoing, down-to-earth quality manager who leads by example, has a friendly demeanor, and gets people to follow him or her without difficulty. [VI.B.1]

10. c; Someone who demonstrates low dominance and sociability is a *reflective* communicator. These individuals express themselves in a more deliberate manner, may seem more quiet or introverted, and make decisions more slowly. An example of a reflective communicator is a research scientist known for his or her lab work who prefers the ability to summarize and write white papers without dedicated due dates. [VI.B.1]

11. b; A director communicator ranks high on the dominance scale and low on the sociability scale. A stern, serious, indifferent attitude and strong opinions are typical characteristics of this type of communicator. An example of a director communicator is a company president who is very businesslike, expresses his or her opinions with conviction, and appears to be inflexible. [VI.B.1]

12. d; Avoiding cultural misunderstandings during written and presentation forms of communication is easier because the sender usually has much more time to develop the material before the delivery of the information. [VI.B.1]

13. a; A person from North America would be most likely to have a sequential and low-context communication style. [VI.B.1]

14. d; A person from Asia would be most likely to have a high-context and synchronic communication style. [VI.B.1]

15. b; A Pareto chart can be used to help a process improvement team focus their efforts based on the PFMEA. [VI.B.2]

16. c; The purpose of the Pareto chart is to separate the "vital few" causes from the "trivial many." This is often reflected in what is called the 80/20 rule, and helps focus attention on the more pressing issues. [VI.B.2]

17. b; This figure demonstrates negative correlation. The points fall on a straight line descending from left to right.

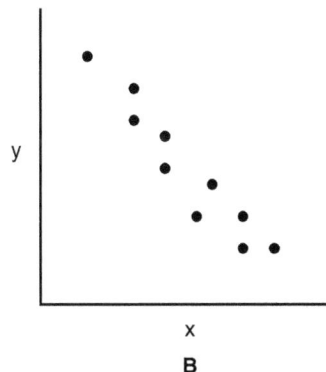

y

x

B

[VI.B.2]

18. d; This figure demonstrates nonlinear correlation (quadratic correlation). There are times when there can be linear, quadratic, cubic, or no correlation present.

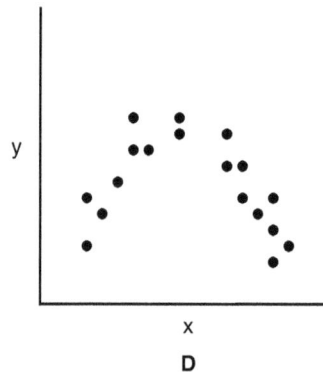

D

[VI.B.2]

19. c; This figure has no discernable relationship. Therefore, there is no correlation present between the two variables.

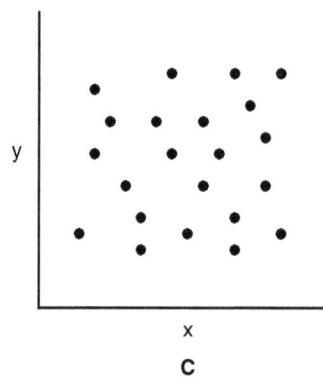

C

[VI.B.2]

20. a; This figure demonstrates positive correlation. The points fall on a straight line ascending from left to right. However, the correlation is not perfect due to the distance between the points (dispersion).

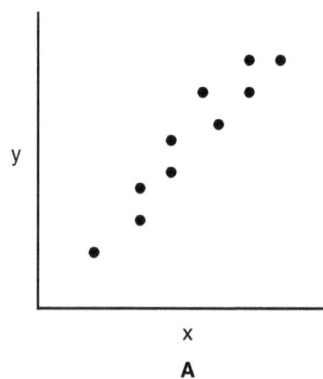

A

[VI.B.2]

21. c; A cause-and-effect diagram (also called an *Ishikawa diagram* or *fishbone diagram*) traditionally divides causes into several generic categories. In use, a large empty diagram is often drawn on a whiteboard or flip chart to visually display potential causes of a problem. [VI.B.2]

22. b; A key benefit of the SIPOC is that it is much easier to complete than either a process map or a value stream map. SIPOCs can be used as a basis for constructing detailed process maps and value stream maps at future dates. Furthermore, SIPOCs help identify the voice of the customer as well as provide quick oversight into the initial X's and Y's. [VI.B.2]

23. c;

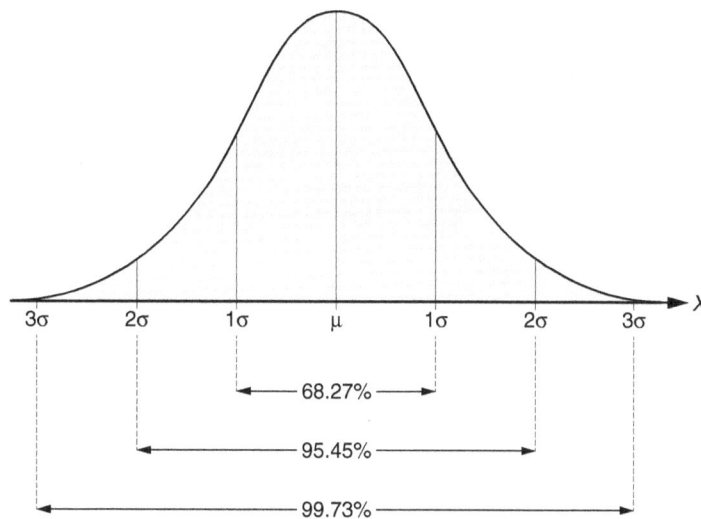

The percentages for ±1, ±2 , and ±3 standard deviations are provided in the figure. [VI.B.2]

24. a; Once a process has been improved, it must be monitored to ensure that the gains are maintained and to determine when additional improvements are required. Control charts are used to monitor the stability of the process and determine when a special cause is present, and when to take appropriate action. [VI.B.2]

25. c; Control charts generally contain a centerline (process average) and upper and lower control limits (±3σ). Specification limits are generally not plotted or recorded on statistical control charts. [VI.B.2]

26. c; Check sheets are used to collect data in real time at the location where the data are generated. Check sheets can use quantitative or qualitative data. [VI.B.2]

27. d; Flowcharts are graphical representations of the steps in a process. Flowcharts are drawn to better understand processes. [VI.B.2]

Part VI
Answers

28. b; A cause-and-effect diagram (also called the *Ishikawa diagram* or *fishbone diagram*) traditionally divides causes into several generic categories. In use, a large empty diagram is often drawn on a whiteboard or flip chart to visually display potential causes of a problem. [VI.B.2]

29. b; Total quality management (TQM) is an organization-wide quality program that focuses on the customers' involvement in determining quality, employees' obligation to quality, and strong leadership from upper management in quality improvement efforts. [VI.C]

30. a; Armand V. Feigenbaum lists three steps to quality: quality leadership, modern quality technology, and organizational commitment. [VI.C]

31. c; The RACI (responsible, accountable, consulted, and informed) model. The RACI matrix is usually defined in detail by a team or committee, and in some cases by an individual. The roles are:

    • *Responsible.* Individuals who actively participate in an activity.

    • *Accountable.* The individual ultimately accountable for results. Only one individual may be accountable at a time.

    • *Consulted.* Individuals who must be consulted before a decision is made.

    • *Informed.* Individuals who must be informed about a decision because they are affected. These individuals do not need to take part in the decision-making process.

    [VI.C]

32. d; The RACI (responsible, accountable, consulted, and informed) model. The RACI matrix is usually defined in detail by a team or committee, and in some cases by an individual. The roles are:

    • *Responsible.* Individuals who actively participate in an activity.

    • *Accountable.* The individual ultimately accountable for results. Only one individual may be accountable at a time.

    • *Consulted.* Individuals who must be consulted before a decision is made.

    • *Informed.* Individuals who must be informed about a decision because they are affected. These individuals do not need to take part in the decision-making process.

    [VI.C]

33. b; The RACI (responsible, accountable, consulted, and informed) model. The RACI matrix is usually defined in detail by a team or committee, and in some cases by an individual. The roles are:

    - *Responsible.* Individuals who actively participate in an activity.

    - *Accountable.* The individual ultimately accountable for results. Only one individual may be accountable at a time.

    - *Consulted.* Individuals who must be consulted before a decision is made.

    - *Informed.* Individuals who must be informed about a decision because they are affected. These individuals do not need to take part in the decision-making process.

    [VI.C]

34. a; The RACI (responsible, accountable, consulted, and informed) model. The RACI matrix is usually defined in detail by a team or committee, and in some cases by an individual. The roles are:

    - *Responsible.* Individuals who actively participate in an activity.

    - *Accountable.* The individual ultimately accountable for results. Only one individual may be accountable at a time.

    - *Consulted.* Individuals who must be consulted before a decision is made.

    - *Informed.* Individuals who must be informed about a decision because they are affected. These individuals do not need to take part in the decision-making process.

    [VI.C]

35. d; Directional leadership is characterized by high directive and low supportive behavior. It is suited for individuals who possess low competence and high commitment. Such individuals are often described as the *enthusiastic beginner.* They are new to a task and motivated about it. [VI.C]

36. b; Supportive leadership style is characterized by low directive and high supportive behavior. It is suited for individuals who possess moderate competence and commitment. Such individuals are often described as the *capable, but cautious, performer.* They know their task, but are not motivated about it. [VI.C]

37. a; Delegative leadership style is characterized by low directive and low supportive behavior. It is suited for individuals who possess high competence and high commitment. Such individuals are often described as the *self-reliant.* [VI.C]

Part VII

Business Governance, Ethics and Compliance

(20 questions)

A. ASQ CODE OF ETHICS

Determine appropriate behavior in situations requiring ethical decisions, including identifying conflicts of interest, recognizing and resolving ethical issues. (Apply)

B. COMPLIANCE

Understand issues of compliance and their applicable policies, laws and regulations (e.g., conflict of interest, confidentiality, bribery). (Apply)

C. CONFIDENTIALITY

1. *Organizational policies.* Apply organizational policies for executing appropriate agreements such as, non-disclosure, quality, and change notification agreements. (Apply)

2. *Intellectual property.* Apply procedures for protecting the intellectual property of an organization and its suppliers. (Apply)

3. *Illegal activity.* Understand and interpret policies for reporting observations and deviations that could be perceived as illegal activity. (Apply)

QUESTIONS

1. A situation that exists when an individual exploits, or appears to exploit, their position for personal gain is a(n):

   a. emotional conflict.

   b. conflict of conscience.

   c. conflict of interest.

   d. conflict of effort.

2. The ASQ Code of Ethics applies to:

   a. ASQ members.

   b. ASQ certification holders.

   c. ASQ members and ASQ certification holders.

   d. all quality professionals.

3. The ASQ Code of Ethics requires ASQ members and ASQ certification holders during relations with the public to:

   a. perform services in any area of quality.

   b. hold paramount the safety, health, and welfare of the public in the performance of their professional duties.

   c. assure billing practices are fair.

   d. perform services for only one client at a time.

4. Supervisory independence and the use of non-executive board members is a:

   a. socially responsible investment (SRI).

   b. method to help minimize ethical issues.

   c. case for competitors as stakeholders.

   d. requirement of the ASQ Code of Ethics.

Use the following case study to answer questions 5–8 below.

> *You are performing a supplier qualification audit. The audit is not going well, but it's clear the supplier wants the business very badly, as does your R&D function. It's also clear the supplier can see it's going poorly. The general*

manager and quality assurance manager see gaps that your experience can fill. They ask you to help them to close those gaps. The gaps are serious but not critical or safety related. They are not a sole supplier or sole source.

5. Is it okay to help with those gaps?

    a. No. An auditor must remain objective. After the audit, if all parties see value in you helping, it could be acceptable.

    b. Yes, because they are trying their best and have not offered money or any incentive.

    c. Yes, because R&D will say supplier quality is setting up obstacles again.

    d. No, because it is a clear conflict of interest.

6. How should you handle this situation?

    a. End the audit. This supplier is not qualified and will fail the audit anyway.

    b. Explain that you are flattered but you must remain objective, and continue.

    c. Help them set up an action plan as you go forward so they have each finding covered.

    d. Ignore the request as addressing it will only increase the awkwardness.

7. At the conclusion of the audit you should tell your manager:

    a. about the incident, how you handled it, and why.

    b. nothing. It is not a big deal, you did the right thing. It'll call your skills into question.

    c. how unqualified they are. End the issue by not qualifying the supplier.

    d. you'll try to institute damage control with R&D, or they will blame you for not qualifying their supplier.

8. How would this scenario change if the supplier *did* offer you pay for consulting?

    a. It doesn't. Handle it the same way.

    b. It now becomes a higher level of conflict of interest because they are trying to incentivize you.

    c. It's still okay as long as you consult on your free time.

    d. It doesn't. It's not a bribe to alter the audit report or anything dishonest.

9. Supplier quality agreements should include:

   a. a change notification process.

   b. a nondisclosure agreement/IP protections.

   c. applicable quality system requirements.

   d. all of the above.

10. A statement of work (SOW):

   a. is most appropriate for a service provider (lab, consultant).

   b. can be used interchangeably for service and manufacturing suppliers.

   c. provides details on expectations and deliverables.

   d. both (a) and (c).

11. Purchasing terms:

   a. are appropriate to include in a quality agreement.

   b. should be kept apart from a quality agreement.

   c. are handled by supplier quality professionals.

   d. dictate the type of quality agreement.

12. Supplier-related organizational policies:

   a. are best implemented at onboarding.

   b. protect corporate information.

   c. define requirements.

   d. all of the above.

13. In the US, intellectual property (IP) violations may be reported to:

   a. the US Patent Office.

   b. the US Attorney's Office.

   c. the National Intellectual Property Rights Coordination Center.

   d. your member of Congress.

14. WIPO is an organization that:

   a. protects IP rights globally by administering treaties.

   b. polices violations of IP.

    c. offers registration of IP.

    d. coordinates with the US Patent Office.

15. Intellectual property can be protected:

    a. by lawsuits and court actions.

    b. only in the US.

    c. by registering with the WIPO Convention.

    d. by applying for patents, copyrights, and trademarks.

16. IP infringements should be handled _____ with the supply base.

    a. on a case-by-case basis

    b. only as needed

    c. within a standard, documented process

    d. swiftly and punitively

17. The difference between illegal activities and conflicts of interest is:

    a. semantics. The real difference is in the quantity of incentives offered.

    b. that laws vary from jurisdiction to jurisdiction, so it's a judgement call.

    c. that someone can go to prison for one and not the other.

    d. that one is a clear violation of a legal statute and the other is an ethical violation.

18. Illegal activities include all but (choose one):

    a. paying for a private practitioner's airfare.

    b. bribes to officials.

    c. trading in illegal materials.

    d. trading in restricted materials (certain types of technology).

19. The EU's REACH and RoHS are examples of:

    a. hazardous materials regulations.

    b. medical device regulations.

    c. food safety regulations.

    d. employment laws.

20. US law provides protection to "whistleblowers" under the:

    a. Taft-Hartley Amendments.

    b. Sarbanes-Oxley Act.

    c. Dodd-Frank Act.

    d. Wagner Act.

ANSWERS

1. c; A conflict of interest occurs when an individual acts in a way that exploits a given situation to better themselves. [VII.A]

2. c; The ASQ Code of Ethics Fundamental Principles apply to ASQ members and ASQ certification holders. [VII.A]

3. b; The ASQ Code of Ethics requires ASQ members and ASQ certification holders in Article 1 to hold paramount the safety, health, and welfare of the public in the performance of their professional duties in relation with the public. [VII.A]

4. b; Supervisory independence and the use of non-executive board members are methods to help minimize ethical issues. [VII.A]

5. a; This is a conflict of interest. Objectivity is the auditor's currency. However, since they might become a business partner if all parties agree and the audit results stand objectively, it could be acceptable after the audit concluded. The other answers show a lack of understanding of ethics and conflicts of interest. [VII.B]

6. b; There is no need to end the audit at this point. Remain objective and polite but carry on. The other answers show a lack of understanding of ethics and conflicts of interest. [VII.B]

7. a; You must report the offer—it could be indicative of future issues. Hiding it and burying it won't help, and could be perceived as dishonest. The other answers show a lack of understanding of ethics and conflicts of interest. [VII.B]

8. b; The situation is now serious, and crossed any line of possible misunderstanding the minute money was offered. The other answers show a lack of understanding of ethics and conflicts of interest. [VII.B]

9. d; All of these elements are typically found in a quality agreement. [VII.C.1]

10. d; Although it's not impossible to fit an SOW to a manufacturing supplier, it would not be the best format choice. (a) and (c) are found in typical SOWs. [VII.C.1]

11. b; Purchasing terms should be kept apart from a quality agreement. [VII.C.1]

12. d; Supplier-related organizational policies are best implemented at onboarding, help protect corporate information, and define requirements. [VII.C.1]

13. c; US ICE runs this office and any enforcement will flow from reporting to the national office. The US Attorney might enforce after reporting. The Patent Office and Congress provide the means and the laws for registration of IP. [VII.C.2]

14. a; WIPO is only an organization that administers global treaties. These other activities are for individual nations to carry out. [VII.C.2]

15. d; Registration comes first. Enforcement comes after. WIPO provides international protection through treaties ([b] is wrong therefore), not registration. [VII.C.2]

16. c; IP needs to be handled up front and documented with suppliers in a uniform way to prevent issues later. [VII.C.2]

17. d; (c) is true but not the most accurate distinction. (a) and (b) are false statements. [VII.C.3]

18. a; (b), (c), and (d) are all clearly illegal. [VII.C.3]

19. a; Both REACH and RoHS are environmental statutes regulating the import and use of hazardous materials. [VII.C.3]

20. c; Whistleblowers are protected under Dodd-Frank. (d) is a labor law dealing largely with unions and overtime, (a) is an amendment to the Wagner Act, and Sarbanes-Oxley is a finance law enacted in the aftermath of the Enron scandal. [VII.C.3]

Section 2
Practice Test

(150 questions)

QUESTIONS

1. A situation that exists when an individual exploits, or appears to exploit, their position for personal gain is a(n):

    a. emotional conflict.

    b. conflict of conscience.

    c. conflict of interest.

    d. conflict of effort.

2. Which stage of the PDCA cycle would start with a spend analysis calculation?

    a. Plan

    b. Do

    c. Check

    d. Act

3. Which of the following ISO standards provides guidance on auditing management systems?

    a. ISO 9001

    b. ISO 13485

    c ISO 15378

    d. ISO 19011

4. Which phase of the supplier orientation process is used to gather work requirements and methods, prepare work requirements for suppliers per scope of work, and establish expectations for suppliers?

   a. Prototyping

   b. Planning

   c. Development

   d. Execution

5. The primary difference between supplier life cycle management (SLM) and supply chain management (SCM) is:

   a. SLM focuses on supplier integration.

   b. SLM's assessment of suppliers' assets and capabilities.

   c. SCM's planning for turnover in suppliers.

   d. SCM focuses on supplier integration.

6. This audit team role assists the lead auditor in organizing the team and making it more effective, but does not participate as a team member.

   a. Client

   b. Auditor

   c. Facilitator

   d. Lead auditor

7. The _____ shows the relationship between requirements, risks, and needs of the customer.

   a. demand planning

   b. house of quality

   c. PFMEA

   d. design output

8. Documentation of supplier selection activities must be maintained to ensure _____ over time.

   a. record control

   b. the supplier file

   c. consistency

   d. fairness

9. When using GD&T, MMC refers to:

   a. median material condition.

   b. mean material condition.

   c. minimum material condition.

   d. maximum material condition.

10. Which of the following tools would be the most appropriate for crisis management?

    a. PEST analysis

    b. SWOT analysis

    c. Contingency planning

    d. Risk analysis

11. Which of the following European Union (EU) regulations restricts the use of certain chemicals, including lead (Pb), mercury (Hg), and cadmium (Cd), in electrical and electronic products?

    a. REACH

    b. RoHS

    c. OSHA

    d. Conflict Minerals

12. Drawing dimensions can be linear, circular, or:

    a. rectangular.

    b. triangular.

    c. angular.

    d. singular.

13. Which of the following tools is typically used first in a lean system?

    a. 5S

    b. Kanban

    c. Value stream map

    d. Visual control

14. Supplier remediation activities include:

    a. PPAP instruction.

    b. joint kaizen events.

    c. technical assessments.

    d. all of the above.

15. The costs associated with the operation and activities of the material review board (MRB) are:

    a. appraisal costs.

    b. prevention costs.

    c. internal failure costs.

    d. external failure costs.

16. Which of the following is a characteristic of a team that is effectively functioning?

    a. A sense of team interdependence

    b. A clear sense of purpose

    c. A focus on individual tasks

    d. A knowledge of company policies

17. Postproduction risk management can be derived from various sources, including service personnel, training personnel, incident reports, and _____.

    a. production monitoring.

    b. process validation.

    c. customer feedback.

    d. incoming inspection.

18. A value indicating the relative risk of a potential failure is referred to as:

    a. failure.

    b. modality.

    c. RPN.

    d. severity.

19. An FMEA has a severity of 7, a probability of occurrence of 5, and a probability of detection of 3. What is the RPN?

    a. 12

    b. 15

    c. 35

    d. 105

20. A systematic approach that proactively identifies, analyzes, prioritizes, and documents potential failure modes and their respective potential causes of failures is:

    a. DFSS.

    b. FMEA.

    c. SIPOC.

    d. PDCA.

21. A visual identification of many potential causes of a problem is referred to as a:

    a. process flowchart.

    b. cause-and-effect diagram.

    c. decision tree.

    d. check sheet.

22. Internal stakeholders for requirements include:

    a. R&D/engineering.

    b. procurement/purchasing.

    c. operations.

    d. all of the above.

23. Deviations from specifications might be acceptable if:

    a. engineering signs off on *use as-is* status.

    b. procurement gets the supplier to lower their piece part price sufficiently to pay for more incoming inspection.

    c. there is sufficient engineering/scientific objective evidence that there will be no adverse effect on form/fit/function, performance, or safety.

    d. there is sufficient need to launch the product on time and within budget.

24. The purchasing/procurement function has two fundamental duties:

    a. to select and contract with suppliers and set terms of purchased goods and services.

    b. to ensure competitive bids and select suppliers based on cost and delivery.

    c. to source suppliers at the lowest cost and seek cost reductions going forward.

    d. to select the best supplier and negotiate a fair price.

25. This type of communicator expresses themself in a more deliberate manner, may seem more quiet or introverted, and makes decisions more slowly.

    a. Emotive

    b. Director

    c. Reflective

    d. Supportive

26. SMART goals—This method helps ensure that the goals have been fully investigated and provides a way to clearly understand the implications of the goal-setting process. What does the M refer to?

    a. Meaningful

    b. Management

    c. Metrics

    d. Measurable

27. The purpose of risk evaluation is to:

    a. determine options for modifying risks.

    b. compare risk level to stated risk criteria.

    c. decide whether risk levels are acceptable.

    d. evaluate and possibly change the consequences.

28. The APQP process is designed to be used during:

    a. the product development process.

    b. the RFQ process.

    c. the supplier qualification process.

    d. the product validation process.

29. The acronym COPQ represents:

    a. cost of perfect quality.

    b. cost of product quality.

    c. cost of present quality.

    d. cost of poor quality.

30. The ASQ Code of Ethics requires ASQ members and ASQ certification holders during relations with the public to:

    a. perform services in any area of quality.

    b. hold paramount the safety, health, and welfare of the public in the performance of their professional duties.

    c. assure billing practices are fair.

    d. perform services for only one client at a time.

31. When determining a make/buy decision, which tool would be best for analyzing historical performance?

    a. DMAIC

    b. PDCA

    c. CAPA

    d. FMEA

32. The team phase where team members begin to understand the need to operate as a team rather than as a group of individuals is referred to as:

    a. forming.

    b. storming.

    c. norming.

    d. performing.

33. Other requirements in a technical review can include:

    a. manufacturing environment (for example, cleanroom).

    b. functional specifications.

    c. material properties.

    d. all of the above.

34. A team was put together to analyze potential actions that could lead to process failures. The goal is for the team to help determine the risk associated with process failures and to prevent future failures from occurring. Which quality tool would help the team visualize this scenario?

    a. Fault tree

    b. Affinity diagram

    c. Matrix diagram

    d. Process decision program chart

35. The type of leader who provides clear direction to team members is:

    a. supportive.

    b. coaching.

    c. situational.

    d. directional.

36. The net present value (NPV) costs to conduct a project are estimated to be $100,000. The NPV benefits or savings due to the project are estimated at $750,000. Compute the benefit-to-cost ratio.

    a. $0.13

    b. $7.50

    c. $100,000

    d. $750,000

37. The causal factor or factors that, if removed, will prevent the recurrence of the same situation describe:

    a. root cause analysis.

    b. FMEA.

    c. process control.

    d. root cause.

Use the following figure to answer question 38.

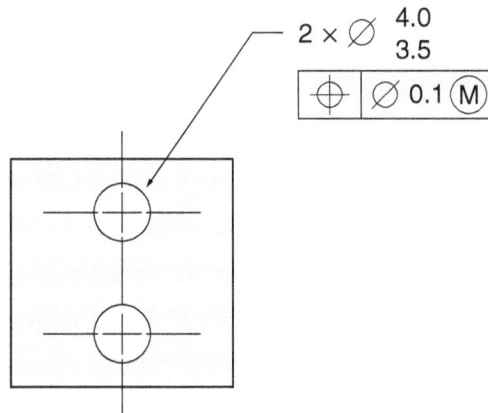

38. What is the bonus tolerance if the feature is 3.7?

    a. 0.0

    b. 0.2

    c. 0.3

    d. 0.5

39. A technique that can be used by supplier quality engineers for the generation of ideas—where ideas are written down without any discussion, after which the ideas are ranked—is:

    a. brainstorming.

    b. affinity analysis.

    c. nominal group technique.

    d. configuration management.

40. Available data do not always provide a reliable basis for predicting the future. For unique types of risks or for new product types, _____ data may not be available.

    a. attribute

    b. variables

    c. neutral

    d. historical

For questions 41–43, refer to the following table. Higher values are more desirable.

Supplier	Price	Weight	Quality	Weight	Capacity	Weight	Totals
A	9	1	6	1	3	2	21
B	3	1	6	1	9	2	27
C	6	1	9	1	3	2	21
D	6	1	9	1	6	2	27

41. Based on the table, what's the most important category for comparison?

    a. Price

    b. Quality

    c. Capacity

    d. All are equal

42. Which supplier is most desirable?

    a. A

    b. B

    c. C

    d. D

43. Which approach does this table represent?

    a. FMEA RPN

    b. Control plan

    c. The "best ranking approach"

    d. Risk–benefit analysis

44. Which phase of the supplier orientation process is used to communicate the requirements, means and methods, schedules, and deliverables per scope of work?

    a. Prototyping

    b. Planning

    c. Development

    d. Execution

45. The costs associated with the implementation of a companywide training initiative are:

    a. appraisal costs.

    b. prevention costs.

    c. internal failure costs.

    d. external failure costs.

46. This audit team role is responsible for calling the meetings, making meeting arrangements, running the meetings, and reporting progress to the client and the auditee.

    a. Client

    b. Team members

    c. Facilitator

    d. Lead auditor

47. An engineer needs to copy an ISO A2 drawing. What sheet size (in millimeters) does the engineer need?

    a. 841×1189

    b. 594×841

    c. 420×594

    d. 297×420

Use the following information to answer questions 48–50.

A supplier quality engineer is completing the quarterly scorecard for a high-risk supplier using the following criteria.

Product quality:

$$\frac{\text{Total number of line items} - \text{Number of SNCRs}}{\text{Total number of line items}} \times 60$$

On-time delivery:

$$\frac{\text{Number of on-time deliveries}}{\text{Total number of line items}} \times 40$$

Scorecard rating = Product quality + On-time delivery

Total line items for Q3: 25 items

Total SCARS: 3

Number of on-time deliveries: 20

48. The product quality score is:

    a. 32.0.

    b. 40.0.

    c. 52.8.

    d. 60.0.

49. The on-time delivery score is:

    a. 32.0.

    b. 40.0.

    c. 52.8.

    d. 60.0.

50. The supplier Q3 scorecard rating is:

    a. 40.0.

    b. 60.0.

    c. 84.8.

    d. 100.0.

51. A process flow diagram (PFD) does which of the following?

    a. Controls the process

    b. Diagrams a lean material system using the principle of flow

    c. Maps the process steps and sequence of operations in a process

    d. Displays where product goes in the value stream

52. The need for facilitators and facilitation techniques _____ with team member training, experience, and successful team history.

    a. increases

    b. decreases

    c. remains the same

    d. is not considered

53. Strategic products are characterized as products with

    a. low profit impact and low supply risk.

    b. low profit impact and high supply risk.

    c. high profit impact and low supply risk.

    d. high profit impact and high supply risk.

54. Quality risk management supports a(n) _____ approach to decision making.

    a. anecdotal

    b. biased

    c. procedural

    d. scientific

55. IP infringements should be handled _____ with the supply base.

    a. on a case-by-case basis

    b. only as needed

    c. within a standard, documented process

    d. swiftly and punitively

56. Supplier performance is optimized long term by:

    a. open communication.

    b. collaboration and planning.

    c. joint problem solving.

    d. all of the above.

57. Which statement is an example of a vision statement?

    a. To procure the highest-quality components from the most reliable suppliers at the best cost possible

    b. To comply with all applicable standards and regulations

    c. To complete all supplier audits per the specified schedule

    d. To create a reliable supply chain through unwavering commitment to quality and cost

Choose from the following list to answer questions 58–60.

A. A step-by-step review of inputs and outputs that relate to a specific operation, or combination of operations

B. A review of the implementation regulatory requirements and adherence

C. A review of elements governing processes, policies, and records for the assurance of quality

D. A review of critical-to-quality characteristics to understand performance measures such as out-of-the-box failure

58. A process audit is:

a. A.

b. B.

c. C.

d. D.

59. A QMS audit is:

a. A.

b. B.

c. C.

d. D.

60. A product audit is:

a. A.

b. B.

c. C.

d. D.

61. Before suppliers provide a quotation for a new project, what should the supplier consider?

a. Capability analysis

b. Capacity analysis

c. Feasibility analysis

d. Takt time analysis

62. A person from North America would most likely have the following communication style:

    a. low-context and sequential.

    b. low-context and synchronic.

    c. high-context and sequential.

    d. high-context and synchronic.

63. PPAP is required in all of the following situations *except*:

    a. new parts.

    b. pricing changes.

    c. changes in part processing.

    d. sub-tier supplier or materials change.

64. Which of the following is another name for poka-yoke?

    a. Waste

    b. Error-proofing

    c. Orderliness

    d. Pull system

Use the table below to answer questions 65 and 66.

	Supplier A	Supplier B	Supplier C	Supplier D
Part cost	$1750	$1650	$1835	$1800
Performance index (PI)	1.3	1.4	1.1	1.2

65. Which supplier will deliver the lowest elevated cost per part?

    a. Supplier A

    b. Supplier B

    c. Supplier C

    d. Supplier D

66. Which supplier will deliver the highest elevated cost per part?

    a. Supplier A

    b. Supplier B

    c. Supplier C

    d. Supplier D

67. What type of leadership is best suited for high-performing team members that are technically competent?

    a. Delegating

    b. Directional

    c. Supportive

    d. Adversarial

68. Labeling, warning, identification, traceability, risk management, and recall requirements are all part of an effective:

    a. product marketing campaign.

    b. supplier quality agreement.

    c. compliance program.

    d. risk mitigation program.

Use the following case study to answer questions 69 and 70.

> *You are performing a supplier qualification audit. The audit is not going well, but it's clear the supplier wants the business very badly, as does your R&D function. It's also clear the supplier can see it's going poorly. The general manager and quality assurance manager see gaps that your experience can fill. They ask you to help them to close those gaps. The gaps are serious but not critical or safety related. They are not a sole supplier or sole source.*

69. How should you handle this situation?

    a. End the audit. This supplier is not qualified and will fail the audit anyway.

    b. Explain that you are flattered but you must remain objective and continue.

    c. Help them set up an action plan as you go forward so they have each finding covered.

    d. Ignore the request as addressing it will only increase the awkwardness.

70. At the conclusion of the audit you should tell your manager:

    a. about the incident, how you handled it and why.

    b. nothing. It is not a big deal, you did the right thing. It'll call your skills into question.

c. how unqualified they are. End the issue by not qualifying the supplier.

d. you'll try to institute damage control with R&D, or they will blame you for not qualifying their supplier.

Use the following information to answer questions 71–73.

An audit is being performed, specifically focusing on documentation (SOPs and WIs) and records. The documentation policy requires that the "same function" that originally approved the document must approve changes to documents. The procedure also requires that the documents be available to the employees electronically. The records policy requires the record to be legible, protected from unintended alterations, and with electronic scans acceptable as the permanent record. During the audit, the auditor observed the following:

SOP 110 Preventive Maintenance—*the procedure was originally approved 10/1/2017 by Jane Doe, Maintenance Manager, and Bill Smith, Quality Manager.*

SOP 115 Production Control—*the procedure was originally approved 11/12/2016 by Todd Jones, Production Manager, and Bill Smith, Quality Manager.*

Production Batch Record 1234 was scanned and saved on the company server, which is cloud based. The server requires a unique user name and password for access. The batch was approved for release on 3/14/2018.

71. *SOP 110 Preventive Maintenance* needs to be updated; however, Jane Doe, Maintenance Manager, was promoted to director of operations, and a replacement was recently hired. Which of the following individual(s) should approve the change to the procedure?

a. Only the quality manager because he was an original signer

b. The quality manager and the maintenance manager

c. The VP of operations and the maintenance manager

d. The VP of operations, the maintenance manager, and the quality manager

72. *SOP 115 Production Control Documentation* needs to be updated; however, Todd Jones, Production Manager, is on vacation for six weeks. The production manager directly reports to the VP of operations. Which of the following individual(s) should approve the change to the procedure?

a. The VP of operations and the quality manager.

b. The VP of operations and the maintenance manager.

   c. Only the quality manager because he was an original signer.

   d. The procedure should not be changed until production manager Todd Jones returns from vacation.

73. The auditor noticed that *Production Batch Record 1234* was approved for release on 3/14/2018, scanned, and saved to the cloud-based server. The auditor asked for the original paper copy. The VP of operations informed the auditor that under the procedure scanned electronic copies are acceptable as the permanent record. Is this a violation of the procedure?

   a. Yes, original paper documents should never be destroyed.

   b. Yes, the cloud-based server is not physically located on the company property.

   c. No, paper copies are required to be saved.

   d. No, the procedure allows electronic scans as the permanent record.

74. Audit reports should not include:

   a. argumentative statements.

   b. subjective opinions.

   c. proprietary information.

   d. all of the above.

75. With the globalization of the supply chain, supplier partnerships are increasingly necessary to:

   a. comply with regulations.

   b. deliver quality products.

   c. ensure market share.

   d. provide export licenses.

76. Analyzing the problem is associated with which phase of the PDCA cycle?

   a. Plan

   b. Do

   c. Check

   d. Act

77. Which quality tool is used to diagram a process?

    a. Control chart

    b. Measles chart

    c. Venn diagram

    d. Flowchart

78. A customer order for 1000 quarters is due in five days. Company ABC runs one eight-hour shift per day. What is the takt time?

    a. .417

    b. 2.4

    c. 25

    d. 625

79. In the 5S methodology, the Japanese word for *sorting* is:

    a. Seiri

    b. Seiton

    c. Seiso

    d. Seiketsu

80. The EU's REACH and RoHS are examples of:

    a. hazardous materials regulations.

    b. medical device regulations.

    c. food safety regulations.

    d. employment laws.

81. Excess motion, repairs/rework, waiting, and excess inventory are some examples of:

    a. steps in a process.

    b. muda.

    c. cycle time reduction.

    d. steps to take to ensure material is ready when production is.

82. Process validation consists of:

    a. running at rate, RFQ, process optimization.

b. control plans, PSW, process map.

c. material performance tests, biocompatibility, and sterilization.

d. IQ, OQ, and PQ.

83. A supplier is using a pull system for managing inventory. Inventory management pull systems are also known as:

    a. hoshin.

    b. kaizen.

    c. kanban.

    d. SMED.

Use the figure below to answer questions 84 and 85.

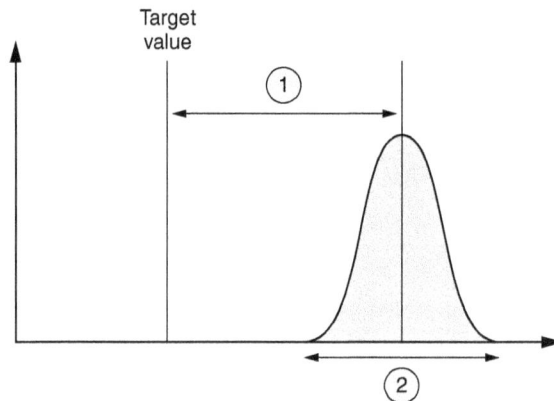

84. The encircled number 1 relates most closely to:

    a. the lower control limit.

    b. the upper control limit.

    c. accuracy.

    d. precision.

85. The encircled number 2 relates most closely to:

    a. the lower control limit.

    b. the upper control limit.

    c. accuracy.

    d. precision.

86. The production part approval process (PPAP):

    a. has no value outside of the automotive industry.

    b. establishes that components and subassemblies meet requirements.

    c. is too difficult for most suppliers.

    d. includes the elements of quoting, costing, and demand planning.

87. Lean manufacturing principles are the foundation of any successful manufacturing company. The implementation of these principles can be done in any order, as long as the implementation is effective and sustainable. Which of the following is least likely to be used for a lean initiative?

    a. 5S

    b. Kanban

    c. Standard work

    d. Design of experiments

88. A team facilitator can promote positive team relationships by:

    a. recording meeting notes.

    b. debating each item.

    c. keeping track of time.

    d. encouraging participation.

89. The need for a supplier typically arises in the _____ phase.

    a. product development

    b. procurement

    c. production

    d. participation

90. Intellectual property can be protected:

    a. by lawsuits and court action.

    b. only in the US.

    c. by registering with the WIPO Convention.

    d. by applying for patents, copyrights, and trademarks.

91. What type of plan is used to identify unrealistic time and cost estimates, customer review cycle, budget cuts, changing requirements, and lack of committed resources?

   a. Risk management

   b. Communication

   c. Budget

   d. Project

92. Gage R&R studies are important because:

   a. they establish measurement system error and, if acceptable, ensure that the data we collect for other acceptance activities are accurate and precise.

   b. these studies prove through objective evidence that an inspector is proficient.

   c. these studies qualify inspectors to complete PPAP.

   d. these studies highlight the differences in technique between inspectors.

Use the following information to answer questions 93–96.

> A lot containing 100 printed circuit boards (PCBs) was received and sent to incoming inspection. The inspection plan called for 15 randomly selected boards to be tested. During incoming inspection, the 15 randomly selected PCBs failed the "burn-in" test. The 15 PCBs were immediately labeled, placed on electronic hold, and moved to the quarantine storage area. A nonconformance report was initiated and placed on the agenda for the material review board.

93. What is the first action the MRB should perform?

   a. Issue a SCAR to the supplier

   b. Ensure that the remaining 85 PCBs are inspected

   c. Issue a rework order

   d. Ensure that the remaining 85 PCBs are quarantined

94. The PCBs are critical to the product being manufactured. During the MRB meeting, the incoming inspection manager reported that the last lot of this particular PCB was not inspected due to the skip-lot sampling scheme. What action should the MRB perform?

   a. Issue a SCAR to the supplier

   b. Release the remaining 85 PCBs to production

   c. Place all PCBs from the previous lot on hold

   d. Issue a rework order

95. The production manager is worried that the month's production quota will not be met and wants the MRB to release the PCBs from both lots to production and to ship the equipment that was assembled using the PCBs from the skip-lot. What is the most reasonable action for the MRB?

   a. Issue a SCAR to the supplier.

   b. Nothing, the MRB should wait for the supplier's response.

   c. Perform a retest based on an AQL sampling on the equipment that was assembled using the PCBs from the skip-lot and release.

   d. Perform a retest based on all of the equipment that was assembled using the PCBs from the skip-lot and release.

96. The supplier of the PCBs informs the MRB that a new manufacturer of a component was utilized beginning with the lot of 100 PCBs that failed incoming inspection. What is the most reasonable action for the MRB?

   a. Issue a SCAR to the supplier.

   b. Nothing, the MRB should wait for the supplier's response.

   c. Perform a retest based on an AQL sampling on the equipment that was assembled using the PCBs from the skip-lot and release.

   d. Perform a retest based on all of the equipment that was assembled using the PCBs from the skip-lot and release.

97. This type of specification defines what is acceptable as raw material entering a manufacturing process.

   a. Quality management specifications

   b. Product specifications

   c. Analytical specifications

   d. Raw material specifications

98. Which of the following models is the best for identifying stakeholders of a proposed change in the supply chain?

   a. SIPOC

   b. PDCA

    c. DMAIC

    d. DFSS

99. _____ is a very important part of the orientation process.

    a. Cooperation

    b. Feedback

    c. Collaboration

    d. Good communication

100. In which role in the RACI model must the individual be cognizant about a decision because they are affected?

    a. Responsible

    b. Accountable

    c. Consulted

    d. Informed

Use the following figure to answer question 101.

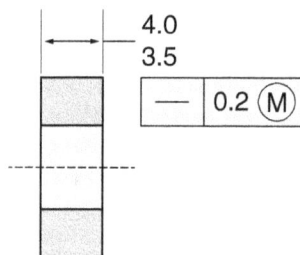

101. What is the bonus tolerance if the feature is 3.7?

    a. 0.0

    b. 0.2

    c. 0.3

    d. 0.5

102. The primary purpose of the control plan is to:

    a. keep only one item as a variable and maintain all other characteristics as controls per experiment.

    b. implement validations on specific processes.

    c. ensure that process changes are maintained over time.

    d. create structure for how quality will be implemented throughout the process.

103. Which quality guru developed the *kanban* concept?

    a. Genichi Taguchi

    b. Masaaki Imai

    c. Taiichi Ohno

    d. W. Edwards Deming

104. The tool that can be used to organize the ideas developed during a brainstorming session into larger categories is the:

    a. interrelationship diagram.

    b. Gantt chart.

    c. affinity diagram.

    d. work breakdown structure.

105. _____ should be used to prioritize supplier development opportunities, escalation/exit strategy, and future sourcing preferences.

    a. Pricing and schedule fulfillment

    b. Performance metrics

    c. On-time delivery scores

    d. Willingness to reduce cost year over year

106. This is a situation in which critical information is withheld from the team because individual members censor or restrain themselves, either because they believe their concerns are not worth discussing or because they are afraid of confrontation.

    a. Storming

    b. Socializing

    c. Brainstorming

    d. Groupthink

107. It is a best practice when concluding a technical review to:

    a. thank the supplier.

    b. review the items discussed.

    c. document the agreed-on output of the meeting.

    d. rely on memory.

108. What to do when the critical parameter goes out of control is indicated on the control plan. It would most likely be found in the _____ column.

    a. reaction plan

    b. characteristic

    c. specification

    d. frequency

109. A tool used to monitor a process to make sure that the process improvements made are continuing to produce desired results is:

    a. the control plan.

    b. process metrics.

    c. control charts.

    d. gage R&R.

110. When developing standard work, the scope of work should be outlined or documented on a:

    a. check sheet.

    b. kanban card.

    c. flowchart.

    d. control plan.

111. Purchasing terms:

    a. are appropriate to include in a quality agreement.

    b. should be kept apart from a quality agreement.

    c. are handled by supplier quality professionals.

    d. dictate the type of quality agreement.

112. In-process acceptance is calculated by part acceptance rates. Acme Manufacturing, a supplier of key components, provided the following information for Q4: 768 parts were manufactured, of which 17 parts were found to be incorrect and had to be discarded. What is the in-process acceptance percentage rate for Q4?

    a. 2.21

    b. 45.18

c. 97.79

d. 100.00

113. A PFMEA is a _____ tool.

    a. risk management

    b. quality management

    c. sales forecasting

    d. quality engineering

114. Which of the following is considered by many experts to be the sixth S in a 5S program?

    a. Safety

    b. Sort

    c. Shine

    d. Sustain

115. During which stage of the PDCA cycle would supply chain management program development and procurement activities be started with a few selected suppliers?

    a. Plan

    b. Do

    c. Check

    d. Act

116. The phrase "vital few and useful many" is applicable to the:

    a. cause-and-effect diagram.

    b. check sheet.

    c. Pareto diagram.

    d. process flow diagram.

Questions 117 through 121 concern types of supplier assessment metrics. Choose the category based on the list offered.

117. Percent on-time delivery and percent late deliveries are examples of:

    a. timeline metrics.

    b. quality metrics.

    c. delivery (JIT) metrics.

    d. compliance metrics.

118. Percent over-orders, percent under-ordered, and early delivery are examples of:

    a. timeline metrics.

    b. quality metrics.

    c. delivery (JIT) metrics.

    d. compliance metrics.

119. Percent supplies missing certificates and percent of reported required quality information are examples of:

    a. timeline metrics.

    b. quality metrics.

    c. delivery (JIT) metrics.

    d. compliance metrics.

120. Percent defective, PPM, DPMO, and percent nonconforming are examples of:

    a. cost metrics.

    b. quality metrics.

    c. delivery (JIT) metrics.

    d. compliance metrics.

121. Dollars rejected versus purchased are examples of:

    a. cost metrics.

    b. quality metrics.

    c. delivery (JIT) metrics.

    d. compliance metrics.

122. How can a team best overcome anxiety during the forming stage?

    a. Create a team name and slogan

    b. Let members pick their own teams

    c. Determine a team leader quickly

    d. Set clear goals, timelines, and roles to communicate expectations

123. Supplier selection activities can all be boiled down to _____ to choose the best overall supplier.

    a. characterizing comparative risks of several factors

    b. determining the lowest cost and highest quality

    c. understanding the supplier's comparative capacity and lowest cost

    d. determining the supplier with the highest comparative capability and (potential) service levels

124. Single-minute exchange of die (SMED) generally has a target time of _____ minutes.

    a. 1

    b. 10

    c. 30

    d. 60

125. A person from Asia would most likely have what type of communication style?

    a. Low-context and sequential

    b. Low-context and synchronic

    c. High-context and sequential

    d. High-context and synchronic

126. Registration to meet the EU requirements has been adopted to protect human health and the environment and to prevent risk posed by chemicals. This registration is known as:

    a. REACH

    b. RoHS

    c. OSHA

    d. Conflict Minerals

127. Residual risk is:

    a. risk that exists before a risk treatment has been implemented.

    b. risk due to a gap between what is thought to be risk and what really is risk.

    c. unknown risk that can never be identified using FMEA.

    d. risk that remains after a risk treatment has been implemented.

128. The best person to approve a supplier audit response is:

    a. the supplier quality manager.

    b. the director of procurement.

    c. the supplier's management team.

    d. the auditor.

129. Which of the following cases would be considered the least desirable in terms of the relationship of the specification to the process spread?

    a. $6\sigma < USL - LSL$

    b. $6\sigma = USL - LSL$

    c. $6\sigma > USL - LSL$

    d. $3\sigma = USL$

130. US law provides protection to "whistleblowers" under the:

    a. Taft-Hartley Amendments

    b. Sarbanes-Oxley Act

    c. Dodd-Frank Act

    d. Wagner Act

131. What are the five stages of team development, in order?

    a. Performing, storming, norming, adjourning, forming

    b. Storming, adjourning, norming, forming, performing

    c. Forming, storming, norming, performing, adjourning

    d. Storming, performing, forming, norming, adjourning

132. An often overlooked component of supplier qualification is:

    a. the audit.

    b. sub-tier suppliers.

    c. financial due diligence.

    d. the supplier questionnaire.

133. Incoming acceptance can be done through either lot or part acceptance. Acme Manufacturing, a supplier of key components, shipped 27 lots in Q1, of which one was rejected. What is the percentage of lot acceptance for Q1?

    a. 3.70

    b. 3.85

    c. 92.30

    d. 100.00

134. A tool used to provide visual identification of many potential causes of a problem is the:

    a. decision tree.

    b. process flowchart.

    c. cause-and-effect diagram.

    d. check sheet.

135. The tool that captures requirements on inputs into and outputs from a process is:

    a. requirements tree analysis.

    b. SIPOC.

    c. quality function deployment.

    d. the Kano model.

136. According to the Kraljic portfolio segmentation model, a product or service with low profit impact, high supply risk, and high sourcing difficulty would be considered:

    a. leverage products.

    b. strategic products.

    c. routine products.

    d. bottleneck products.

137. The costs associated with acceptance sampling are:

    a. appraisal costs.

    b. prevention costs.

    c. internal failure costs.

    d. external failure costs.

138. This audit team role ensures that everyone understands his or her assignment before the audit starts.

    a. Client

    b. Auditor

    c. Facilitator

    d. Lead auditor

139. The hierarchy of risk control should follow which of the following sequences?

    a. Substitute, eliminate, engineering controls, administrative controls

    b. Substitute, eliminate, administrative controls, engineering controls

    c. Eliminate, substitute, engineering controls, administrative controls

    d. Eliminate, substitute, administrative controls, engineering controls

140. During design and development, _____ must consider functional and performance requirements, regulatory requirements, and safety requirements.

    a. design and development planning

    b. design and development outputs

    c. design and development changes

    d. design and development inputs

141. The objective of finalizing requirements collaboratively is:

    a. to simultaneously meet goals for quality, cost, and delivery.

    b. to make sure all parties agree.

    c. so the supplier can't say later, "I didn't know."

    d. to make sure purchasing meets their spend reduction targets.

142. _____ needs to be an integral part of the risk treatment plan to ensure that the measures remain effective.

    a. Mitigation

    b. Reporting

    c. Monitoring

    d. Planning

143. Standardizing the solution is associated with which phase of the PDCA cycle?

    a. Plan

    b. Do

    c. Check

    d. Act

144. Critical elements of PPAP to review and approve are:

    a. control plan, MSA, PFMEA, PFD, and capability study.

    b. IQ, OQ, and PQ.

    c. PPV, run at rate, and tooling cost.

    d. CTQs, safety program, and employees training program.

145. CLEAR goals—A newer method for setting goals that takes into consideration the environment of today's fast-paced business. What does the L refer to?

    a. Lesser

    b. Larger

    c. Limited

    d. Leveraged

146. One objective of supplier orientation is to:

    a. impart an understanding of the key work processes and requirements.

    b. introduce all the QA personnel and suppliers to each other.

    c. show the supplier your systems.

    d. give the supplier a chance to showcase their business to management.

147. Incremental continuous improvement is also known as:

    a. kaizen.

    b. Six Sigma.

    c. lean enterprise.

    d. plan–do–check–act.

148. The primary difference between mission and vision statements is:

    a. mission statements relate to the future of the company.

    b. vision statements dictate the future state of the company.

    c. mission statements explain the company's reason for existence.

    d. mission and vision statements are essentially the same and are interchangeable.

149. Armand V. Feigenbaum's concept of total quality includes:

    a. quality leadership, modern quality technology, and organizational commitment.

    b. commitment to quality, management commitment, and measurement of potential quality problems.

    c. statistical process control.

    d. benchmarking and reengineering.

150. Supplier quality agreements should include:

    a. a change notification process.

    b. a nondisclosure agreement/IP protections.

    c. applicable quality system requirements.

    d. all of the above.

ANSWERS

1. c; A conflict of interest occurs when an individual acts in a way that exploits a given situation to better themselves. [VII.A]

2. a; During the *plan* stage a spend analysis and diagnosis of the value being purchased are evaluated. [1.B.4]

3. d; ISO 19011:2011 provides guidance on auditing management systems, including the principles of auditing, managing an audit program, and conducting management system audits, as well as guidance on the evaluation of competence of individuals involved in the audit process, including the person managing the audit program, auditors, and audit teams. [V.A.2]

4. c; The phase of the supplier orientation process used to gather work requirements and methods, prepare work requirements for suppliers per scope of work, and establish expectations for suppliers is the *development* phase. [VI.A]

5. d; SLM is an integrated approach that considers the business and quality needs of the organization. SCM is a set of system-focused approaches utilized to efficiently integrate suppliers, manufacturers, warehouses, and stores so that merchandise is produced and distributed in the right quantities, to the right locations, and at the right time in order to minimize systemwide costs while satisfying service level requirements. The primary difference between SLM and SCM is that SCM focuses on supplier integration. [1.B.1]

6. c; A team may have a facilitator who assists the lead auditor in organizing the team and making it more effective, but does not participate as a team member. In the cross-functional team for a specific purpose and the long-term project team, the facilitator or coach is included as needed. The facilitator's role is to help the team resolve issues and reach its purpose or achieve its goals effectively. [V.B.2]

7. b; The house of quality is a graphical representation of these often competing and complex drivers. Demand planning determines how many parts we need, FMEA is a risk tool, and the house of quality can be used to develop design inputs, not outputs. [III.A.1]

8. c; To ensure fairness, consistency is the key. Yes, these activities are records for the supplier file, but that's not the key concept. [III.B.3]

9. d; MMC indicates maximum material condition. For example, the smallest hole or the largest diameter allowed. [III.C.2]

10. c; Contingency planning (also called a *plan B*) is used for crisis management, business continuity, and asset security. [II.A.1]

11. b; The Restriction of Hazardous Substances (RoHS) Directive restricts the types of hazardous materials that can be used in electrical and electronic products. Under RoHS, the following substances are prohibited from entering the market for electrical and electronic products: lead (Pb), mercury (Hg), cadmium (Cd), hexavalent chromium (Cr VI), polybrominated biphenyls (PBB), polybrominated diphenyl ethers (PBDE), and four different phthalates (DEHP, BBP, DBP, and DIBP). [V.C]

12. c; Drawing dimensions can be linear, circular, or angular. [III.C.1]

13. c; The goal of a lean system is to eliminate waste from a process. The first step is to identify opportunities for improvement, which can be found using a value stream map. Value stream maps are a visual representation of all the steps in a process, including information about timing. Value stream maps also identify the value-added and non-value-added activities in a process. [IV.A.3]

14. d; All of these activities are great focuses to resolve performance issues with the supplier. [V.A.4]

15. c; Internal failure costs are costs incurred when a failure occurs in-house, and are usually associated with the cost of scrap and rework. [1.C.1]

16. b; A characteristic of a team that is effectively functioning is having a clear sense of purpose. [V.B.3]

17. c; Postproduction information can be part of established QMS procedures (for example, monitoring and measuring). Manufacturers must establish procedures to collect information from various sources, such as service personnel, training personnel, incident reports, and customer feedback. The other choices are related to preproduction or production activities. [II.A.2]

18. c; The RPN 24 is a value indicating the relative risk of the potential failure. The RPN is the product of the severity, probability of occurrence, and probability of detection. [II.B.1]

19. d; To calculate the RPN for an FMEA with a severity of 7, a probability of occurrence of 5, and a probability of detection of 3, the following formula is used:

$$RPN = S \times O \times D = 7 \times 5 \times 3 = 105$$

where

S = Severity

O = Probability of occurrence

D = Probability of detection

[II.B.1]

20. b; FMEA is a systematic approach that proactively identifies, analyzes, prioritizes, and documents potential failure modes and their respective potential causes of failures. [II.B.1]

21. b; A cause-and-effect diagram (also called the *Ishikawa diagram* or *fishbone diagram*) traditionally divides causes into several generic categories. In use, a large empty diagram is often drawn on a whiteboard or flip chart to visually display potential causes of a problem. [IV.C.1]

22. d; All three of these groups need to be considered internal stakeholders along with supplier quality and any local purchasing or materials planners. [III.A.2]

23. c; Deviations should only be acceptable if there are engineering studies demonstrating that the change will not have adverse or unintended consequences. [III.C.5]

24. a; Although all four answers have more or less merit, answer (a) describes the basic purchasing function most accurately as it exists in most organizations. [1.D]

25. c; Someone who demonstrates low dominance and sociability is a *reflective* communicator. These individuals express themselves in a more deliberate manner, may seem more quiet or introverted, and make decisions more slowly. An example of a reflective communicator is a research scientist known for his or her lab work who prefers the ability to summarize and write white papers without dedicated due dates. [VI.B.1]

26. d;

S	Specific
M	Measurable
A	Attainable
R	Realistic
T	Timely

[V.A.5]

27. b; Risk evaluation is used to make decisions about which risks need to be addressed. [II.A.2]

28. a; APQP is not related to RFQ, but validation might be a part of it. Supplier qualification might be a part of the development process. But (a) is the right answer—it is a quality process to be used during the design and development process. [III.A.1]

29. d; COPQ is the acronym used for the costs of poor quality. COPQ are categorized as internal, external, prevention, and appraisal. [1.C.1]

30. b; The ASQ Code of Ethics in Article 1 requires ASQ members and ASQ certification holders to hold paramount the safety, health, and welfare of the public in the performance of their professional duties in relations with the public. [VII.A]

31. c; Key data sources for analyzing historical performance include SCARs, corrective and preventive action (CAPA), internal nonconformances, historical capability, and results from regulatory, ISO, and customer inspections and audits. Planned or scheduled maintenance, age of equipment, historical downtime, first-pass yield, and rolled throughput yield are also great predictors of success. [1.C.3]

32. c; Team norming is when members begin to understand the need to operate as a team rather than as a group of individuals. [V.B.1]

33. d; All of these should be included. [III.C.1]

34. a; Fault trees provide a visualization of hierarchical relationships of events that lead to failure (or other undesirable outcome). The team here wishes to determine potential actions that can lead to process failures. [IV.C.1]

35. d; Directional leadership is characterized by high directive and low supportive behavior. It is suited for individuals who possess low competence and high commitment. Such individuals are often described as the *enthusiastic beginner*. They are new to a task and motivated about it. [VI.C]

36. b; The net present value (NPV) costs to conduct a project are estimated to be $100,000. The NPV benefits or savings due to the project are estimated at $750,000. The benefit-to-cost ratio can be calculated by

$$\frac{\Sigma \text{ NPV of all benefits anticipated}}{\Sigma \text{ NPV of all costs anticipated}} = \frac{\$750,000}{\$100,000} = \$7.50$$

[1.C.2]

37. d; Solving a process problem means identifying the root cause and eliminating it. The ultimate test of whether the root cause has been eliminated is the ability to toggle the problem on and off by removing and reintroducing the root cause. [IV.C.1]

38. b; Bonus tolerance = Absolute difference between MMC and actual condition.

$$BT = 3.5 - 3.7 = .02$$

[III.C.2]

39. a; Brainstorming is a group process used to generate ideas in a nonjudgmental way. The purpose of brainstorming is to generate a large number of ideas about an issue. [IV.C.1]

40. d; Available data do not always provide a reliable basis for predicting the future. For unique types of risks or for new product types, historical data may not be available or there may be different interpretations of available data by different stakeholders (industry and regulators, for example). [II.A.3]

41. c; Since capacity is weighted higher, it would be the most critical category. [III.B.1]

42. b; Since there is a numerical tie, and capacity is weighted higher, supplier B would be the choice. [III.B.1]

43. c; "The best ranking approach" uses a table to compare and tabulate numerical rankings based on the most important categories. [III.B.1]

44. d; The *execution* phase of the supplier orientation process is used to communicate the requirements, means and methods, schedules, and deliverables per scope of work. [VI.A]

45. b; Prevention costs are the costs of all activities whose purpose is to prevent failures, including training, quality planning, and quality control activities. [1.C.1]

46. d; The lead auditor is responsible for calling the meetings, making meeting arrangements, running the meetings, and reporting progress to the client and the auditee. The lead auditor should take ownership of the audit process. The lead auditor normally controls the team output or deliverables. Lead auditors must be able to guide the team and make decisions as needed to ensure that the team is effective. The lead auditor should be a qualified (competent) auditor. [V.B.2]

47. c; An ISO A2 drawing sheet size is 420mm × 594mm. [III.C.1]

48. c; Product quality = [(25 − 3) / 25) × 60] = 52.8. [IV.A.1]

49. a; On-time delivery = [(20 / 25) × 40] = 32.0. [IV.A.1]

50. c; Scorecard rating = 52.8 + 32.0 = 84.8. [IV.A.1]

51. c; A PFD is a diagram of all the steps in the process. [III.C.3]

52. b; The need for facilitators and facilitation techniques decreases with team member training, experience, and successful team history. [V.B.2]

53. d; Strategic products (high profit impact, high supply risk). These products deserve the most attention from purchasing managers. Options include developing long-term supply relationships, regularly analyzing and managing

risks, planning for contingencies, and producing the item in-house rather than buying it, if appropriate. [1.B.3]

54. d; Quality risk management supports a scientific approach to decision making. [II.A.3]

55. c; IP needs to be handled up front and documented with suppliers in a uniform way to prevent issues later. [VII.C.2]

56. d; All of these activities are critical to maintaining a healthy level of supplier performance. [1.D]

57. d; A vision statement is what or where the company would like to be in the future. The other answers are more task oriented and do not reflect the future state. [1.A]

58. a; A focus on inputs, outputs, and steps is an indication that a process is being audited. [III.B.2]

59. c; A focus on quality assurance and policies is an indication that a QMS is being audited. [III.B.2]

60. d; Critical-to-quality and out-of-box should lead the reader to understand that a product audit is under way. [III.B.2]

61. c; Before suppliers provide a quotation for a new project, the supplier should consider performing a feasibility analysis. Capability analysis, capacity analysis, and takt time analysis are elements of a feasibility analysis. [1.E]

62. a; A person from North America would most likely have a sequential and low-context communication style. [VI.B.1]

63. b; (a), (c), and (d) are all conditions where PPAP should be performed. Pricing alone is not. [III.C.4]

64. b; Poka-yoke comes from Japan and is also known as *error-proofing*. It is a method of preventive action—a technique used to prevent errors from occurring. [IV.A.3]

65. c; Supplier C has the lowest elevated cost per part ($2018.50).

	Supplier A	Supplier B	Supplier C	Supplier D
Part cost	$1750	$1650	$1835	$1800
Performance index (PI)	1.3	1.4	1.1	1.2
	2275	2310	2018.50	2160

[1.B.2]

66. b; Supplier B has the highest elevated cost per part ($2310).

	Supplier A	Supplier B	Supplier C	Supplier D
Part cost	$1750	$1650	$1835	$1800
Performance index (PI)	1.3	1.4	1.1	1.2
	2275	2310	2018.50	2160

[1.B.2]

67. a; A delegative leadership style is characterized by low directive and low supportive behavior. It is suited for individuals who possess high competence and high commitment. Such individuals are often described as the *self-reliant*. [VI.C]

68. d; Labeling, warning, identification, traceability, risk management, and recall requirements are all part of an effective risk mitigation program. [II.B.2]

69. b; There is no need to end the audit at this point. Remain objective and polite but carry on. The other answers show a lack of understanding of ethics and conflicts of interest. [VII.B]

70. a; You must report the offer—it could be indicative of future issues. Hiding it and burying it won't help, and could be perceived as dishonest. The other answers show a lack of understanding of ethics and conflicts of interest. [VII.B]

71. b; *SOP 110 Preventive Maintenance* needs to be updated; however, Jane Doe, Maintenance Manager, was promoted to director of operations, and a replacement was recently hired. The procedure was originally approved 10/1/2017 by maintenance manager Jane Doe and Bill Smith, Quality Manager. The procedure should be signed by the same function, not necessarily the same individual. Therefore, the maintenance manager and the quality manager should both approve the changes to the procedure. [V.A.1]

72. a; *SOP 115 Production Control Documentation* needs to be updated; however, Todd Jones, Production Manager, is on vacation for six weeks. Because the production manager directly reports to the VP of operations, it would be appropriate for the VP of operations and the quality manager to approve the changes to the document. [V.A.1]

73. d; The auditor noticed that *Production Batch Record 1234* was approved for release on 3/14/2018, scanned, and saved to the cloud-based server. The auditor asked for the original paper copy. The VP of operations informed the auditor that under the procedure, scanned electronic copies are acceptable as the permanent record. This would not constitute a violation of the procedure. However, if the scan copy were not legible, it could be a violation of the procedure. [V.A.1]

74. d; None of these are acceptable in an audit report. Auditors must strive for objectivity and to be factual. Since the report might be read by others, confidentiality (proprietary information) must be safeguarded. [V.A.2]

75. b; While complying with regulations, delivering quality products, ensuring market share, and providing escort licenses all support supplier partnerships, the primary reason for supplier partnerships is to consistently deliver quality products. [1.A]

76. a; Analyzing the problem is associated with the *plan* phase of the PDCA cycle. [II.B.3]

77. d; Flowcharts are graphical representations of the steps in a process. Flowcharts are drawn to better understand processes. [VI.B.2]

78. b; The takt time to produce 1000 quarters due in five days on one eight-hour shift per day is calculated by:

$$\text{Takt time} = \frac{\text{Time available}}{\text{Number of units to be produced}}$$

$$= \frac{8 \times 60 \times 5}{1000} = 2.4 \text{ minutes per unit}$$

[IV.A.3]

79. a; The 5S elements are:

- *Seiri* (Sort). Eliminate whatever is not needed

- *Seiton* (Straighten). Organize whatever remains

- *Seiso* (Shine). Clean the work area

- *Seiketsu* (Standardize). Schedule regular cleaning and maintenance

- *Shitsuke* (Sustain). Make 5S a way of life

[IV.A.3]

80. a; Both REACH and RoHS are environmental statutes regulating the import and use of hazardous materials. [VII.C.3]

81. b; These are verbatim examples of waste, or *muda*. The other answers are simply wrong and would show a need to conceptually revisit not only muda but the other answers as well, if selected. [V.A.4]

82. d; IQ, OQ, and PQ are specific to process validation for biomedical/pharma products. [III.C.4]

83. c; *Kanban* is a Japanese word that means "visual signal" or "card." The signal is to perform a specific task or activity in a certain quantity. This technique is used in production and business processes that rely on a pull or signal to initiate the work. [IV.A.2]

84. c; *Accuracy* refers to how closely the values are centered relative to the target/ nominal valve. [IV.C.2]

85. d; *Precision* refers to how closely the values are grouped or clustered together. [IV.C.2]

86. b; PPAP is a series of quality tools designed to show that requirements are met via GR&R, capability studies, SPC, and PFMEA. [III.C.3]

87. d; Lean manufacturing principles generally do not include the use of design of experiments techniques. [IV.A.2]

88. d; By encouraging participation, the team facilitator can promote positive team relationships. [V.B.3]

89. a; The need for a supplier typically arises in the product development phase. The earlier the need for a supplier is determined, the more easily a supplier can be selected and integrated into the design, development, and production phases. [1.E]

90. d; Registration comes first. Enforcement comes after. WIPO provides international protection through treaties ([b] is wrong therefore), not registration. [VII.C.2]

91. a; A risk management plan can be used to identify all potential risks. These include unrealistic time and cost estimates, customer review cycle, budget cuts, changing requirements, and lack of committed resources. [V.A.5]

92. a; GR&R is critical to establish the validity of future data because it determines error in the entire measurement system: gages, staging and fixtures, and operators (inspectors). [III.C.3]

93. d; The 15 randomly selected PCBs failed the "burn-in" test. These 15 PCBs were immediately labeled, placed on electronic hold, and moved to the quarantine storage area. The MRB should first ensure that the remaining 85 PCBs are placed in quarantine per procedure. [IV.B]

94. c; Because these PCBs are critical to the product being manufactured, the MRB should place all PCBs from the previous lot on hold. [IV.B]

95. d; The best course of action for the MRB would be to perform a retest based on all of the equipment that was assembled using the PCBs from the skip-lot and release. [IV.B]

96. c; Because there is an assignable cause associated with the lot of PCBs that failed incoming inspection, the most reasonable action for the MRB would be to retest based on an AQL sampling on the equipment that was assembled using the PCBs from the skip-lot and release. [IV.B]

97. d; Raw material specifications define what is acceptable as raw material entering a manufacturing process. [V.C]

98. a; Suppliers–inputs–process–outputs–customers (SIPOC) can be used to identify stakeholders. [1.B.1]

99. b; (a), (c), and (d) are all important attitudes and behaviors during the orientation process, but feedback is an actual step in the process. [VI.A]

100. d; The RACI (responsible, accountable, consulted, and informed) model. The RACI matrix is usually defined in detail by a team or committee, and in some cases by an individual. The roles are:

    • *Responsible.* Individuals who actively participate in an activity.

    • *Accountable.* The individual ultimately accountable for results. Only one individual may be accountable at a time.

    • *Consulted.* Individuals who must be consulted before a decision is made.

    • *Informed.* Individuals who must be informed about a decision because they are affected. These individuals do not need to take part in the decision-making process.

    [VI.C]

101. c; Bonus tolerance = Absolute difference between MMC and actual condition.

$$BT = 4.0 - 3.7 = 0.3.$$

    [III.C.2]

102. d; A control plan is a living document that identifies critical input and output variables and associated activities that must be performed to maintain control of the variation of processes, products, and services in order to minimize deviation from their preferred values. A control plan is defined as a living document; it is designed to be maintained. [IV.C.2]

103. c; Taiichi Ohno, an industrial engineer at Toyota, developed kanban to improve manufacturing efficiency. [IV.A.3]

104. c; The affinity diagram is a tool used to organize information and help achieve order out of the chaos that can develop in a brainstorming session. Large amounts of data, concepts, and ideas are grouped based on their natural relationships to one another. [IV.C.1]

105. b; (a) and (c) are important but usually a subset of performance, and without a full picture of performance, price reductions can be very dangerous and increase risk. [1.D]

106. d; Groupthink is a situation in which critical information is withheld from the team because individual members censor or restrain themselves, either because they believe their concerns are not worth discussing or because they are afraid of confrontation. [V.B.3]

107. c; It is *always* a best practice to document results of a supplier meeting, but more so as part of technical review. [III.C.1]

108. a; The control plan should list the steps to be taken when a process change has been detected. This serves as an aid to the responsible personnel during what is often a stressful time. The reaction plan section should cover requirements for containment and inspection of products suspected of having defects. It should also discuss disposition of parts found to be defective. Some control plans prescribe a more intense sampling protocol after certain corrective actions have been taken. [IV.C.2]

109. c; Once a process has been improved, it must be monitored to ensure that the gains are maintained and to determine when additional improvements are required. Control charts are used to monitor the stability of the process, determine when a special cause is present, and when to take appropriate action. [IV.C.2]

110. c; The scope of work should be outlined or documented on a flowchart. [IV.A.2]

111. b; Purchasing terms should be kept apart from a quality agreement. [VII.C.1]

112. c;

$$\frac{768 \ (\text{Total parts}) - 17 \ (\text{Rejected parts})}{768 \ (\text{Total parts})} = 97.79\%$$

[IV.A.1]

113. a; A PFMEA is a tool characterizing risks and their mitigations following the PFD. [III.C.3]

114. a; Many experts consider *safety* to be the sixth S in a 5S program. [IV.A.2]

115. b; Supply chain management program development and procurement activities should be started with a few selected suppliers during the *do* stage. [1.B.4]

116. c; The purpose of the Pareto chart is to separate the "vital few" causes from the "trivial many." This is often reflected in what is called the 80/20 rule, and helps focus attention on the more pressing issues. [VI.B.2]

117. a; Although timeliness and delivery are similar, these are timeliness metrics. They deal with receipts of shipments. [V.A.3]

118. c; Although timeliness and delivery are similar, these are delivery metrics. They deal with just-in-time. [V.A.3]

119. d; Although quality and compliance are similar, these are compliance metrics because they refer to required documentation. [V.A.3]

120. b; Although quality and compliance are similar, these are compliance metrics because they refer to product quality and meeting/not meeting requirements. [V.A.3]

121. a; Because dollar value is mentioned, cost is the right choice. [V.A.3]

122. d; Setting clear goals, timelines, and roles can help a team overcome anxiety during the forming stage. [V.B.1]

123. a; "Overall" is the operative word, so it is about risk over many categories. [III.B.3]

124. b; Single-minute exchange of die (SMED) generally has a target time of 10 minutes. [IV.A.2]

125. d; A person from Asia would most likely have a high-context and synchronic communication style. [VI.B.1]

126. a; Registration, Evaluation, Authorization, and Restriction of Chemicals (REACH) has been adopted to protect human health and the environment and to prevent risk posed by chemicals. [V.C]

127. d; Residual risk is that risk remaining even after a treatment of a known risk has been implemented. [II.A.1]

128. d; Only the auditor has the in-depth first-person knowledge to appropriately judge the acceptability of an audit finding response from the supplier. [V.A.2]

129. a; The least desirable relationship between the specification and the process spread is when $6\sigma <$ USL − LSL, as demonstrated in the figure below.

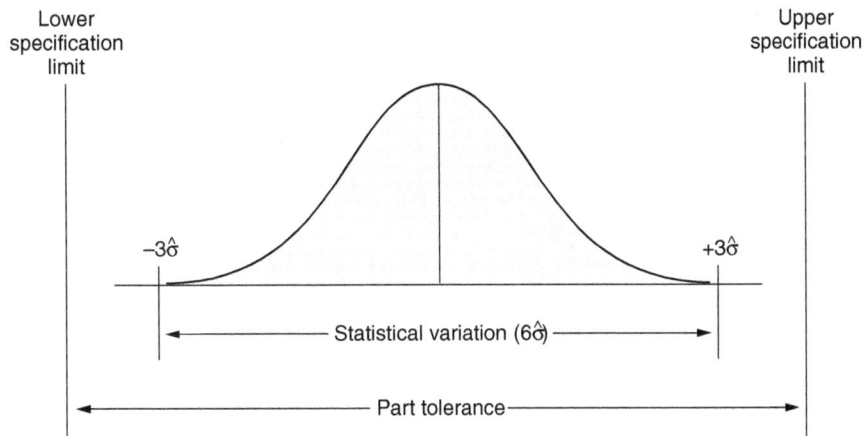

Relationship between statistical control limits and product specifications.

Source: Adapted from Durivage (2014). Used with permission.

[1.C.3]

130. c; Whistleblowers are protected under Dodd-Frank. (d) is a labor law dealing largely with unions and overtime, (a) is an amendment to the Wagner Act, and Sarbanes-Oxley is a finance law enacted in the aftermath of the Enron scandal. [VII.C.3]

131. c; The five stages of team development (in order) are forming, storming, norming, performing, and adjourning. [V.B.1]

132. b; Sub-tier suppliers can be critical to quality. Understand the level and nature of outsourcing by your suppliers. Financial assessment is rarely forgotten and (a) and (d) are very similar. [III.A.2]

133. c;

$$\text{Acceptance } \% = \frac{26 \text{ (Lots accepted)}}{27 \text{ (Lots received)}} = 96.30\%$$

[IV.A.1]

134. c; A cause-and-effect diagram (also called an *Ishikawa diagram* or *fishbone diagram*) traditionally divides causes into several generic categories. In use, a large empty diagram is often drawn on a whiteboard or flip chart to visually display potential causes of a problem. [VI.B.2]

135. b; A key benefit of the SIPOC is that it is much easier to complete than either a process map or a value stream map. SIPOCs can be used as a basis for constructing detailed process maps and value stream maps at future dates. Furthermore, SIPOCs help identify the voice of the customer as well as provide quick oversight into the initial *X*'s and *Y*'s. [VI.B.2]

136. d; The Kraljic portfolio segmentation model classifies products with low profit impact, high supply risk, and high sourcing difficulty as *bottleneck products*. [1.B.3]

137. a; *Appraisal costs* are costs associated with the inspection and appraisal processes. [1.C.2]

138. d; The lead auditor ensures that everyone understands his or her assignment before the audit starts. [V.B.2]

139. c; The hierarchy of risk control should be to eliminate, substitute, provide engineering controls, and provide administrative controls. [II.A.2]

140. d; Inputs define the product, and must meet a whole host of requirements as well as satisfy the design intent. [III.A.1]

141. a; Although (b) and (c) are technically true, the real objective is to meet goals for quality, cost, and delivery. (d) is a nonanswer as it has no bearing. [III.A.2]

142. c; Monitoring needs to be an integral part of the risk treatment plan to ensure that the measures remain effective. [II.B.2]

143. d; Standardizing the solution is associated with the *act* phase of the PDCA cycle. [II.B.3]

144. a; IQ, OQ, and PQ are never PPAP elements. PPV and tooling cost are not PPAP elements, but purchasing data. CTQs and safety are great audit elements. [III.C.5]

145. c;

C	Collaborative
L	Limited
E	Emotional
A	Appreciable
R	Refinable

[V.A.5]

146. a; Although (b), (c), and (d) are all important to accomplish during orientation, the real objective is for the supplier to understand processes and requirements. [VI.A]

147. a; Kaizen refers to incremental improvement that never stops. The other choices are great improvement tools, but are not necessarily used for iterative, ongoing improvement. [V.A.4]

148. c; A mission statement explains the company's reason for existence. A vision statement is what or where the company would like to be in the future. [I.A]

149. a; Armand V. Feigenbaum lists three steps to quality: quality leadership, modern quality technology, and organizational commitment. [VI.C]

150. d; All of these elements are typically found in a quality agreement. [VII.C.1]

NOTES

NOTES

NOTES

NOTES

www.ingramcontent.com/pod-product-compliance
Lightning Source LLC
Chambersburg PA
CBHW081105220326
41598CB00038B/7238